MEMOIRS

of the
American Mathematical Society

WITHDRAWN

Number 444

String Path Integral Realization
of Vertex Operator Algebras

Haruo Tsukada

May 1991 • Volume 91 • Number 444 (first of 4 numbers) • ISSN 0065-9266

American Mathematical Society
Providence, Rhode Island

1980 *Mathematics Subject Classification* (1985 *Revision*).
Primary 17B65, 17B67, 28C20, 30F15, 81E30, 81E40.

Library of Congress Cataloging-in-Publication Data

Tsukada, Haruo, 1961–
    String path integral realization of vertex operator algebras/Haruo Tsukada.
    p. cm. – (Memoirs of the American Mathematical Society, ISSN 0065-9266; no. 444)
    May 1991, volume 91, number 444 (first of 4 numbers)
    Includes bibliographical references.
    ISBN 0-8218-2510-0
    1. Lie algebras. 2. Kac-Moody algebras.   3. Superstring theories.   4. Measure theory.
I. Title. II. Series.
QA3.A57   no. 444
[QA252.3]
510 s–dc20   91-2228
[512′.55]   CIP

Subscriptions and orders for publications of the American Mathematical Society should be addressed to American Mathematical Society, Box 1571, Annex Station, Providence, RI 02901-1571. *All orders must be accompanied by payment.* Other correspondence should be addressed to Box 6248, Providence, RI 02940-6248.

SUBSCRIPTION INFORMATION. The 1991 subscription begins with Number 438 and consists of six mailings, each containing one or more numbers. Subscription prices for 1991 are $270 list, $216 institutional member. A late charge of 10% of the subscription price will be imposed on orders received from nonmembers after January 1 of the subscription year. Subscribers outside the United States and India must pay a postage surcharge of $25; subscribers in India must pay a postage surcharge of $43. Expedited delivery to destinations in North America $30; elsewhere $82. Each number may be ordered separately; *please specify number* when ordering an individual number. For prices and titles of recently released numbers, see the New Publications sections of the NOTICES of the American Mathematical Society.

BACK NUMBER INFORMATION. For back issues see the AMS Catalogue of Publications.

MEMOIRS of the American Mathematical Society (ISSN 0065-9266) is published bimonthly (each volume consisting usually of more than one number) by the American Mathematical Society at 201 Charles Street, Providence, Rhode Island 02904-2213. Second Class postage paid at Providence, Rhode Island 02940-6248. Postmaster: Send address changes to Memoirs of the American Mathematical Society, American Mathematical Society, Box 6248, Providence, RI 02940-6248.

10 9 8 7 6 5 4 3 2 1   95 94 93 92 91

TABLE OF CONTENTS

ABSTRACT

We establish relations between vertex operator algebras in mathematics and string path integrals in physics. In particular, we construct the basic representations of affine Lie algebras of $\widehat{A}\widehat{D}\widehat{E}$-type using a method of string path integrals.

Key words : affine Lie algebra, Virasoro algebra, vertex operator algebra, modular function, string theory, string path integral, zeta-regularized determinant of Laplacian.

INTRODUCTION

Affine Kac-Moody Lie algebras are natural generalizations of finite dimensional simple Lie algebras. They have many important applications such as the Rogers-Ramanujan identities [LW2], soliton equations [DJKM], etc. The simplest class of affine Lie algebras consists of the affine Lie algebras $\widehat{\mathfrak{g}}$ of $\widehat{ADE}$-type. Namely,

$$\widehat{\mathfrak{g}} = \mathfrak{g} \otimes \mathbb{C}[t, t^{-1}] \oplus \mathbb{C}c,$$

where $\mathfrak{g}$ is a simple Lie algebra of ADE-type, and the Lie brackets are given by

$$[x \otimes t^n, y \otimes t^m] = [x, y] \otimes t^{n+m} + <x, y> n\, \delta_{n+m,0} \cdot c,$$

$$[x \otimes t^n, c] = [c, c] = 0,$$

where $< , >$ is a certain symmetric invariant bilinear form on $\mathfrak{g}$ (a normalized Killing form).

In 1978, Lepowsky and Wilson [LW1] constructed the basic representation of $\widehat{sl}(2)$ (the affine Lie algebra of type $\widehat{A}_1$) using operators which turned out to agree with (twisted) vertex operators in string theory in physics (known as dual resonance models at that time). This construction was generalized to the other $\widehat{ADE}$-type affine Lie algebras by Kac, Kazhdan, Lepowsky, and Wilson [KKLW]. In 1980, Frenkel and Kac [FK], and also independently Segal [Seg1] gave another construction of the basic representation of the affine Lie algebras of $\widehat{ADE}$-type by using slight modifications of the (straight) vertex operators in string theory. Later, Borcherds [Bor] used the generalized vertex operators to construct vertex operator algebras associated with even lattices. When the lattice is the root lattice of ADE-type, the vertex operator algebra includes the affine Lie algebras of the corresponding $\widehat{ADE}$-type as subalgebras.

The objective of this paper is to establish relations between vertex operator algebras and string path integrals. In particular, we will construct the basic representations of affine Lie algebras of $\widehat{ADE}$-type by using string path integrals.

A Motivation — Two Approaches to Harmonic Oscillator

There are two major methods in physics, the Hamiltonian method and the Lagrangian method. The construction of vertex operator algebras and the representations of affine

---

Received by the editors March 1989. Revised April 1990.

Research partially supported by a grant from the National Science Foundation.

1

Lie algebras by vertex operators belong to the Hamiltonian method of string theory. On the other hand, string path integrals which we use in this paper belong to the Lagrangian method.

Before considering string theory, let us consider the harmonic oscillator as a motivation. Classically, the potential energy of the harmonic oscillator is given by

$$V(q) = \frac{1}{2}\omega^2 q^2, \quad q \in \mathbb{R},$$

where q is the position and ω is a positive constant. Take the mass $= 1$ for simplicity. The Hamiltonian is

$$H(q,p) = \frac{1}{2}p^2 + \frac{1}{2}\omega^2 q^2,$$

where p is the momentum, and the Lagrangian is

$$L(q,\dot{q}) = \frac{1}{2}\dot{q}^2 - \frac{1}{2}\omega^2 q^2.$$

The equivalence of the two methods is easy. Namely, both methods give the same equation of motion

$$\ddot{q} + \omega^2 q = 0.$$

However, there are significant differences in quantization.

[Hamiltonian method]

In the Hamiltonian approach, we replace p and q with operators P and Q which satisfy the commutation relation

$$[P,Q] = -i.$$

(We have normalized them to remove Planck's constant $\hbar$.) The Hamiltonian becomes

$$H = \frac{1}{2}P^2 + \frac{1}{2}\omega^2 Q^2.$$

Introduce a creation and an annihilation operator by

$$a^* = \frac{1}{\sqrt{2}}(\omega Q - iP) \quad \text{and} \quad a = \frac{1}{\sqrt{2}}(\omega Q + iP).$$

Then we have

$$[a,a^*] = \omega \quad \text{and} \quad H = \frac{1}{2}(aa^* + a^*a) = a^*a + \frac{1}{2}\omega.$$

The operators a^* and a act naturally on the Fock space of polynomials

$$V = \mathbb{C}[a^*]$$

by multiplication and differentiation. ($a = \omega \frac{\partial}{\partial a^*}$.)

The eigenvalues of H are $\{(n + \frac{1}{2})\omega\}_{n=0}^{\infty}$, and the partition function is

$$\text{Tr}_V e^{-TH} = \sum_{n=0}^{\infty} e^{-T(n+\frac{1}{2})\omega} = \frac{1}{e^{\frac{1}{2}\omega T} - e^{-\frac{1}{2}\omega T}}, \quad (T > 0).$$

[Lagrangian method]

In the Lagrangian approach, we consider the Hilbert space

$$\mathcal{H} = L^2(\mathbb{R}, e^{-\omega x^2} dx)$$

of all square-integrable $\mathbb{C}$-valued functions on $\mathbb{R}$ with respect to the Gaussian measure $e^{-\omega x^2} dx$. Passing to imaginary time t (i.e. replacing t by it), the Lagrangian becomes

$$L(f,t) = -\frac{1}{2} \dot{f}(t)^2 - \frac{1}{2} \omega^2 f(t)^2,$$

where f is an $\mathbb{R}$-valued function on the interval $[0, T]$.

We can make sense of a formal expression of a path integral

$$K_T(x,y) = \int_{\substack{f:[0,T]\to\mathbb{R} \\ f(0)=x, f(T)=y}} e^{-I(f)} [df], \qquad I(f) = \int_0^T -L(f,t)\, dt,$$

by using the Wiener measure $d\mu_{x,y}^T$ on the space of continuous paths and we define

$$K_T(x,y) = \int_{\substack{f:[0,T]\to\mathbb{R} \\ f(0)=x, f(T)=y}} \left\{ e^{-\int_0^T \frac{1}{2}\omega^2 f(t)^2\, dt} \right\} d\mu^T(f).$$

We can compute $K_T(x,y)$ explicitly, and we have

$$K_T(x,y) = K_T(0,0)\, e^{-I(f_0)},$$

where

$$K_T(0,0) = \frac{\sqrt{\omega}}{\sqrt{e^{\omega T} - e^{-\omega T}}},$$

and f_0 is the unique function such that

$$(-\Delta + \omega^2) f_0 = 0, \qquad f_0(0) = x, \qquad f_0(T) = y,$$

where $\Delta = \frac{d^2}{dt^2}$ is the Laplacian. Moreover,

$$I(f_0) = \frac{1}{2} \frac{\omega}{e^{\omega T} - e^{-\omega T}} \left\{ (e^{\omega T} + e^{-\omega T})(x^2 + y^2) - 4xy \right\}.$$

The major properties of the path integrals are the followings.

(1) [Feynman-Kac formula]   The Fock space V can be regarded as a subspace of $\mathcal{H}$, and the action of the operator e^{-TH} is realized by

$$\int_{\mathbb{R}} K_T(x,y) F(y)\, e^{-\frac{1}{2}\omega y^2}\, dy = \left(e^{-TH} \cdot F \right)(x)\, e^{-\frac{1}{2}\omega x^2},$$

for all $F \in V$. That is, $K_T(x,y)$ is the integral kernel of the operator e^{-TH}.

(2) [Markov Property, Semigroup Property]

$$\int_{\mathbb{R}} K_T(x,y)\, K_S(y,z)\, dy = K_{T+S}(x,z).$$

(3) We have the following path integral realization of the partition function.

$$\int_{\mathbb{R}} K_T(x,x)\,dx = \int_{\substack{f:[0,T]\to\mathbb{R}\\f(0)=f(T)}} e^{-I(f)}\,[df] = \mathrm{Tr}_V\ e^{-TH}.$$

Although physicists may consider these formulas trivial since both methods describe the same system, they are mathematically very non-trivial. (See [GJ] for details.)

Two Approaches to String Theory

Now we consider the (closed) string moving in the compactified space $\mathbf{S}^1 = \mathbb{R}/2\pi\mathbb{Z}$. As in the case of the harmonic oscillator, there are two approaches to this theory.

Classically, the Hamiltonian is

$$H(x) = \frac{1}{2\pi}\int_0^{2\pi}\left\{\frac{1}{2}\left(\frac{\partial x}{\partial t}\right)^2 + \frac{1}{2}\left(\frac{\partial x}{\partial\theta}\right)^2\right\}d\theta,$$

and the Lagrangian is

$$L(x) = \frac{1}{2\pi}\int_0^{2\pi}\left\{\frac{1}{2}\left(\frac{\partial x}{\partial t}\right)^2 - \frac{1}{2}\left(\frac{\partial x}{\partial\theta}\right)^2\right\}d\theta.$$

Here $\theta \in [0,2\pi]$ parametrizes the string, and $x(t,\theta)$ is the position of the string. We assume that $x(t,0) = x(t,2\pi)$ for all t. As above, we have the equivalence of the both methods and get the same equation of motion

$$\frac{\partial^2 x}{\partial t^2} - \frac{\partial^2 x}{\partial\theta^2} = 0.$$

We quantize the string by analogy with the harmonic oscillator. (See [GSW] for general reference.)

[Hamiltonian method]

Let us take the Fourier expansion

$$x(t,\theta) = q_0(t) + \sum_{n=1}^{\infty} q_n(t)\sqrt{2}\cos(n\theta) + \sum_{n=1}^{\infty} q_{-n}(t)\sqrt{2}\sin(n\theta) + k\theta, \qquad k \in \mathbb{Z}.$$

Then the Hamiltonian is

$$H(x) = \sum_{n\neq 0}\left(\frac{1}{2}\dot{q}_n^2 + \frac{1}{2}n^2 q_n^2\right) + \frac{1}{2}\dot{q}_0^2 + \frac{1}{2}k^2.$$

We get infinitely many harmonic oscillators with creation and annihilation operators such that

$$[a_n, a_n^*] = |n|, \qquad n \neq 0,$$

and a momentum operator P_0 and a position operator K. The Fock space is

$$V = \mathbb{C}[a_1^*, a_2^*, a_3^*, \cdots, a_{-1}^*, a_{-2}^*, a_{-3}^*, \cdots] \otimes L^2(\mathbf{S}^1) \otimes \mathbb{C}[\mathbb{Z}],$$

and a_n^* acts by multiplication, a_n acts as a differentiation $|n|\frac{\partial}{\partial a_n^*}$, $P_0 = -i\frac{d}{d\theta}$ on $L^2(\mathbf{S}^1)$, K acts on $\mathbb{C}[\mathbb{Z}]$ by $K \cdot e^n = n\,e^n$. We regard

$$\mathbb{C}[\mathbb{Z}] \subset \ell^2 = \{\{\lambda_n\}| \sum_{n \in \mathbb{Z}} |\lambda_n|^2 < \infty\}.$$

V is a modification of the Veneziano model in string theory. (For the Veneziano model, see Fubini-Gordon-Veneziano [FGV], Nambu [N], and Susskind [Su]. See also Mandelstam [M].)

Formally the Hamiltonian is

$$H = \frac{1}{2}\sum_{n \neq 0}(a_n^* a_n + a_n a_n^*) + \frac{1}{2}P_0^2 + \frac{1}{2}K^2.$$

The problem is that H is not an operator on V. For example,

$$H \cdot 1 = \frac{1}{2}\sum_{n \neq 0}|n| \cdot 1 = \infty \cdot 1$$

does not converge. (1 is the vacuum state.) One way of interpreting this infinity is the use of an equality

$$\zeta(-1) = -\frac{1}{12},$$

where ζ is the Riemann zeta function which is the analytic continuation of

$$\zeta(s) = \sum_{n=1}^{\infty}\frac{1}{n^s}, \quad \mathrm{Re}(s) > 1.$$

Since formally we have

$$\zeta(-1) = \sum_{n=1}^{\infty} n = 1 + 2 + 3 + \cdots,$$

we define a renormalized Hamiltonian

$$H^{ren} = \sum_{n \neq 0} a_n^* a_n + \frac{1}{2}P_0^2 + \frac{1}{2}K^2 - \frac{1}{12}.$$

This choice of renormalization is particularly good and the partition function is given by

$$\mathrm{Tr}_V \, e^{-TH^{ren}} = \frac{\left(\sum\limits_{n \in \mathbb{Z}} e^{-\frac{n^2}{2}T}\right)^2}{e^{-\frac{1}{12}T}\prod\limits_{n \neq 0}(1 - e^{-|n|T})} = \frac{\theta_{00}(iT/2\pi)^2}{\eta(iT/2\pi)^2},$$

where η is the Dedekind eta function and θ_{00} is the theta function.

[Lagrangian method]

We take the Hilbert space $\mathcal{H}$ of all square-integrable $\mathbb{C}$-valued functionals on a completion $C^\infty(\mathbf{S}^1, \mathbf{S}^1)^\wedge$ of $C^\infty(\mathbf{S}^1, \mathbf{S}^1)$ with respect to a certain Gaussian measure which is formally written as

$$e^{-(f,f)}\, df = \prod_{n \neq 0} e^{-|n|f_n^2}\, df_n\, df_0\, dk,$$

for

$$f(\theta) = f_0 + \sum_{n=1}^\infty f_n \sqrt{2}\cos(n\theta) + \sum_{n=1}^\infty f_{-n} \sqrt{2}\sin(n\theta) + k\theta,$$

where df_n are (normalized) Lebesgue measures on $\mathbb{R}$, df_0 is a (normalized) Lebesgue measure on $\mathbf{S}^1$, and dk is the counting measure on $\mathbb{Z}$.

Passing to imaginary time as before, the Lagrangian is

$$L(\phi, t) = -\frac{1}{2\pi} \int_0^{2\pi} \left\{ \frac{1}{2}\left(\frac{\partial \phi}{\partial t} \right)^2 + \frac{1}{2}\left(\frac{\partial \phi}{\partial \theta} \right)^2 \right\} d\theta.$$

We have to define mathematically a formal expression of a string path integral on the cylinder $C_T = \{ (t, \theta) \mid 0 \leq t \leq T, 0 \leq \theta \leq 2\pi \}$,

$$K_C(f, g) = \int_{\substack{\phi:[0,T]\times\mathbf{S}^1 \to \mathbf{S}^1 \\ \phi(0,\theta)=f(\theta), \phi(T,\theta)=g(\theta)}} e^{-I(\phi)}\, [d\phi].$$

Here the action is

$$I(\phi) = \int_0^T -L(\phi, t)\, dt = \frac{1}{4\pi} \int_0^T \int_0^{2\pi} \left\{ \left(\frac{\partial \phi}{\partial t} \right)^2 + \left(\frac{\partial \phi}{\partial \theta} \right)^2 \right\} d\theta\, dt.$$

Since

$$I(\phi + \phi_0) = I(\phi) + I(\phi_0),$$

when $\phi(0, \theta) = \phi(T, \theta) = 0$ and ϕ_0 is harmonic, we have formally that

$$K_C(f, g) = K_C(0, 0) \sum_{\phi_0} e^{-I(\phi_0)}.$$

The sum is over all harmonic maps ϕ_0 with the boundary condition

$$\phi_0(0, \theta) = f(\theta) \quad \text{and} \quad \phi_0(T, \theta) = g(\theta).$$

Also a formal calculation yields

$$K_C(0, 0) = \int_{\substack{\phi:[0,T]\times\mathbf{S}^1 \to \mathbf{S}^1 \\ \phi(0,\theta)=\phi(T,\theta)=0}} e^{-\frac{1}{4\pi} \iint \phi(-\Delta)\phi\, d\theta\, dt}\, [d\phi] = \frac{1}{\sqrt{\det(-A\Delta)}},$$

where A is the area of the cylinder and

$$\Delta = \frac{\partial}{\partial t^2} + \frac{\partial}{\partial \theta^2}$$

is the Dirichlet Laplacian. Of course the determinant of the Laplacian diverges, and we have to replace $K_C(0,0)$ with a finite constant K_C. One way of defining K_C is the use of zeta-regularized determinant of the Laplacian. (See [RS], [Haw] ,[SchwarzA], [DP], [Fo], [KKW].) This particular choice leads to nice properties listed below. Let $\Lambda = \{\lambda_1, \lambda_2, \lambda_3, \ldots\}$ be the eigenvalues of the operator $-A\Delta$ and we define

$$\zeta_\Lambda(s) = \sum_{n=1}^{\infty} \frac{1}{\lambda_n{}^s}$$

for $\mathrm{Re}(s)$ large enough. This function has a unique analytic continuation to a neighborhood of $s = 0$ ([MP], [See3]) and we define the zeta-regularized determinant by

$$\mathrm{det}_\zeta(-A\Delta) = \left(\prod_{n=1}^{\infty} \lambda_n\right)_\zeta = \exp\left(-\frac{d}{ds}\zeta_\Lambda(0)\right).$$

(This definition generalizes the usual determinant of operators on finite dimensional vector spaces.) Using this, we define a constant

$$K_C = \frac{1}{\sqrt{\mathrm{det}_\zeta(-A\Delta)}}.$$

Note that although the action I is conformally invariant, we have to choose a metric on the cylinder to define the zeta-regularized determinant. Moreover, K_C depends on the choice of a metric (conformal anomaly).

We can also compute the path integral $K_T(0,0)$ using this method. Namely we have

$$K_T(0,0) = \frac{1}{\sqrt{\mathrm{det}_\zeta(-\Delta + \omega^2)}}.$$

We have the following results similar to the harmonic oscillator case.

(1) The Fock space V can be regarded as a subspace of $\mathcal{H}$, and the action of the operator $e^{-TH^{ren}}$ is realized by

$$\int_{C^\infty(\mathbf{S}^1,\mathbf{S}^1)^\wedge} K_{C_T}(f,g)F(g)\,e^{-\frac{1}{2}(g,g)}[dg] = \left(e^{-TH^{ren}} \cdot F\right)(f)\,e^{-\frac{1}{2}(f,f)},$$

for all $F \in V$. That is, $K_{C_T}(f,g)$ is the integral kernel of the operator $e^{-TH^{ren}}$.

(2) [Markov Property, Semigroup Property]

$$\int_{C^\infty(\mathbf{S}^1,\mathbf{S}^1)^\wedge} K_{C_T}(f,g)\,K_{C_s}(g,h)[dg] = K_{C_{T+s}}(f,h).$$

(3) Let $E = E_T$ be the torus which is obtained by joining the two ends of the cylinder C_T together. We define a constant Z_E by

$$Z_E = \frac{1}{\sqrt{\det_\zeta(-A\Delta)}} \sum_{\phi_0} e^{-I(\phi_0)},$$

where A is the area of the torus, Δ is the Laplacian, and the sum is over the harmonic maps of form

$$S_{n,m}(t,\theta) = n\theta + m\frac{2\pi t}{T},$$

where $n, m \in \mathbb{Z}$. We regard Z_E as a string path integral

$$\int_{\phi:E\to\mathbf{S}^1} e^{-I(\phi)}[d\phi].$$

We have the following string path integral realization of the partition function.

$$\int_{C^\infty(\mathbf{S}^1,\mathbf{S}^1)^\wedge} K_{C_T}(f,f)[df] = Z_{E_T} = \mathrm{Tr}_V\, e^{-TH^{ren}}.$$

(The use of symbol $[df]$ is not standard. The precise meaning of the above formulas will be explained in the main text.)

Vertex Operator Algebras

Vertex operator algebras belong to the Hamiltonian approach to string theory. In Chapter I, we review the construction of the vertex operator algebras associated with even lattices ([Bor],[FLM1],[FLM2]).

Let L be an even lattice of rank ℓ. Namely, $L \cong \mathbb{Z}^\ell$ as a group, and L is equipped with a positive biadditive form $< , >$ such that $< \alpha, \alpha >$ is always even for any $\alpha \in L$. The root lattices of ADE-type (with $< \alpha, \alpha >= 2$ for roots α) and the Leech lattice are important examples. We set $\mathfrak{h} = L \otimes_\mathbb{Z} \mathbb{C}$ and extend $< , >$ bilinearly to $\mathfrak{h}$. Define the Heisenberg Lie algebra $\widehat{\mathfrak{h}}$ of $\mathfrak{h}$ by

$$\widehat{\mathfrak{h}} = \mathfrak{h} \otimes \mathbb{C}[t, t^{-1}] \oplus \mathbb{C}c.$$

The Lie brackets are defined as

$$[\alpha(n), \beta(m)] = < \alpha, \beta > n\, \delta_{n+m,0} \cdot c,$$

$$[\alpha(n), c] = [c, c] = 0.$$

Here $\alpha(n)$ means $\alpha \otimes t^n$. This is the Lie algebra of creation and annihilation operators. Take the negative part $\widehat{\mathfrak{h}}^- = \sum_{n=1}^\infty \mathfrak{h} \otimes t^{-n}$ of $\widehat{\mathfrak{h}}$ and we define

$$V_L = Sym(\widehat{\mathfrak{h}}^-) \otimes \mathbb{C}[L],$$

$$V_{L'} = Sym(\widehat{\mathfrak{h}}^-) \otimes \mathbb{C}[L'],$$

$$V_{(\omega)} = Sym(\widehat{\mathfrak{h}}^-) \otimes \mathbb{C}[L + \omega].$$

Here L' is the dual lattice of L, and $\omega \in L'/L$. They are almost half of (ℓ-fold tensor product of) the Fock space of the closed strings we defined earlier.

We define a grading of V_L by the degree

$$\deg(\alpha_1(-n_1) \cdots \alpha_N(-n_N)e^\alpha) = n_1 + \cdots + n_N + \frac{<\alpha, \alpha>}{2}.$$

Then we have

$$V_L = \sum_{n=0}^\infty V_n,$$

where $V_n = \{v \in V_L \,|\, \deg(v) = n\}$.

Using the creation and the annihilation operators $\alpha(n)$, we construct a vertex operator

$$Y(v, z) : V_{L'} \to V_{L'}{}^*,$$

for each element v of V_L and $z \in \mathbf{C} - \{0\}$. Here $V_{L'}{}^*$ is the algebraic dual of $V_{L'}$. There is a certain bilinear form $< \,|\, >$ on $V_{L'}$ induced by $< , >$ on L, and for any $u, v \in V_L$ and $v', v'' \in V_{L'}$, the three series of functions

$$<v'|Y(u, z)Y(v, w)v''> = \sum_{p'} <v'|Y(u, z)p'><p'|Y(v, w)v''>, \quad \text{for } |z| > |w| > 0,$$

$$<v'|Y(v, w)Y(u, z)v''> = \sum_{p'} <v'|Y(v, w)p'><p'|Y(u, z)v''>, \quad \text{for } |w| > |z| > 0,$$

$$<v'|Y(Y(u, z-w)v, w)v''> = \sum_{p} <v'|Y(p, w)v''><p|Y(u, z-w)v>$$

$$\text{for } |w| > |z - w| > 0,$$

converge absolutely. The sums are over a real homogeneous orthonormal basis $\{p\}$ of V_L, and $\{p'\}$ of $V_{L'}$.

Moreover the above three functions of z and w are equal to each other as elements of

$$\mathbf{C}\left[z, \frac{1}{z}, w, \frac{1}{w}, \frac{1}{z-w}\right].$$

This fact is called the commutativity and the associativity of the vertex operators. When v is homogeneous, the vertex operator has an expansion

$$Y(v, z) = \sum_{n \in \mathbf{Z}} v(n)z^{-n-\deg v},$$

where

$$v(n) : V_{L'} \to V_{L'}$$

is a degree $= -n$ operator on $V_{L'}$. The operators $v(n)$, $v \in V_L$, $n \in \mathbf{Z}$ are closed under the Lie bracket and we have

$$[u(n), v(m)] = \sum_{i=1-\deg u}^{\deg v} \binom{n + \deg u - 1}{i + \deg u - 1}(u(i) \cdot v)(n + m),$$

when u and v are homogeneous.

The Lie algebra of the operators $v(n)$ is called the vertex operator algebra of L. (This definition is slightly different from the one in [FLM2], but is essentially the same.) It acts irreducibly on $V_{(\omega)}$.

The vertex operator algebra includes the Virasoro algebra

$$\widehat{\mathcal{L}} = \sum_{n \in \mathbb{Z}} \mathbb{C} L_n \oplus \mathbb{C} c',$$

with the Lie brackets

$$[L_n, L_m] = (n - m)L_{n+m} + \frac{1}{12}(n^3 - n)\delta_{n+m,0} \cdot c',$$

$$[L_n, c'] = [c', c'] = 0,$$

as a subalgebra.

When L is the root lattice of ADE-type, the degree $= 1$ part V_1 of the Fock space V_L is the simple Lie algebra $\mathfrak{g}$ of ADE-type, and the operators $v(n)$ where $v \in \mathfrak{g}$ and $n \in \mathbb{Z}$ define a representation of the affine Lie algebra $\widehat{\mathfrak{g}}$ of the corresponding $\widehat{ADE}$-type. V_L is the (distinguished) basic representation of $\widehat{\mathfrak{g}}$, and $V_{(\omega)}$ are the all level$= 1$ standard representations of $\widehat{\mathfrak{g}}$.

We also define a double Fock space

$$U_{L'} = Sym(\widehat{\mathfrak{h}}^{\pm}) \otimes \mathbb{C}[L'],$$

where $\widehat{\mathfrak{h}}^{\pm} = \sum_{n \neq 0} \mathfrak{h} \otimes t^n$, and construct a neutral vertex operator

$$Y(v, z) : U_{L'} \to U_{L'}{}^*,$$

for any $v \in U_{L'}$. These do not possess a good algebra structure. However we will see later that they have a very nice geometric realization.

Functional Realization of Fock Spaces

Chapters II and III belong to the Lagrangian approach to string theory. These chapters show the relations between the vertex operator algebras and string path integrals.

Let L be an even lattice with a biadditive form $<\ ,\ >$ as before, and we define a new lattice

$$\Gamma = \frac{1}{\sqrt{2}} L.$$

Namely $\Gamma = L$ as groups, but the biadditive form $<\ ,\ >_\Gamma$ of Γ is defined as

$$<\alpha, \beta>_\Gamma = \frac{1}{2} <\alpha, \beta>.$$

Let $\mathbf{T}_L$ be the torus $\mathfrak{h}_\mathbb{R}/2\pi L$, where $\mathfrak{h}_\mathbb{R} = L \otimes_\mathbb{Z} \mathbb{R}$ is a real vector space. We define the Hilbert space $\mathcal{H}_L$ of all square-integrable functionals on a completion $C^\infty(\mathbf{S}^1, \mathbf{T}_L)^\wedge$ of $C^\infty(\mathbf{S}^1, \mathbf{T}_L)$ with respect to a certain Gaussian measure which is formally written as

$$e^{-(f_* \overline{f}_*)} df = \prod_{n=1}^{\infty} e^{-2n<f_n, \overline{f}_n>_\Gamma} df_n d\overline{f}_n \, df_0 \, d\lambda,$$

where

$$f = f_* + f_0 + \lambda\theta, \quad f_*(\theta) = \sum_{n \neq 0} f_n e^{in\theta}, \quad f_n \in \mathfrak{h}, \ f_{-n} = \overline{f}_n, \ f_0 \in \mathbf{T}_L, \ \lambda \in L,$$

$df_n d\overline{f}_n$ are (normalized) Lebesgue measures on $\mathfrak{h}$, df_0 is a (normalized) Lebesgue measure on $\mathbf{T}_L$ and $d\lambda$ is the counting measure on L.

$\mathcal{H}_L$ is the Hilbert space of the closed strings moving in the compactified space $\mathbf{T}_L$ with respect to $< , >_\Gamma$ (not with respect to $< , >$ on L). To construct representation of the vertex operator algebra of L, this Hilbert space has to be modified. We use the Hilbert space $\widetilde{\mathcal{H}}_L$ of the square-integrable functionals F such that

$$F(f + 2\pi\mu) = (-1)^{<\mu,\lambda>} \cdot F(f),$$

for all $f(\theta) = f_* + f_0 + \lambda\theta$ (now $f_0 \in \mathfrak{h}_{\mathbb{R}}$ instead of $\mathbf{T}_L$) and for all $\mu \in L$. We have

$$\widetilde{\mathcal{H}}_L \cong F_0 \cdot \mathcal{H}_L,$$

where $F_0(f) = e^{i<f_0,\lambda>/2}$. The space

$$W = \sum_{\omega \in L'/L} V_{(\omega)} \otimes \overline{V_{(\omega)}}.$$

can be regarded as a subspace of $\widetilde{\mathcal{H}}_L$. ($\overline{V_{(\omega)}}$ is a copy of $V_{(\omega)}$.) Namely, W is isomorphic to the subspace spanned by functionals F such that

$$F(f) = P(f_1, \dots, f_K, \overline{f}_1, \dots, \overline{f}_K) \cdot \delta_{\lambda, r-s} \cdot e^{i<r+s,f_0>/2},$$

for $f(\theta) = \sum_{n\neq 0} f_n e^{in\theta} + f_0 + \lambda\theta$, where P is a polynomial function of finitely many variables $f_1, \dots, f_K, \overline{f}_1, \dots, \overline{f}_K \in \mathfrak{h}$ and $r, s \in L'$, $r - s \in L$.

Also the double Fock space $U_{L'}$ is isomorphic to the subspace spanned by functionals F such that

$$F(f) = P(f_1, \dots, f_K, \overline{f}_1, \dots, \overline{f}_K) \cdot \delta_{\lambda, 0} \cdot e^{i<\mu,f_0>},$$

where $\mu \in L'$.

String Path Integrals

Let X be a Riemann surface with boundary consisting of N circles ($N \geq 1$). Let $f_1, \dots, f_N : \mathbf{S}^1 \to \mathbf{T}_L$ be maps on the components of the boundary. Let $C^\infty{}_{X,L}(f_1, \dots, f_N)$ be the space of maps $\phi : X \to \mathbf{T}_L$ with the boundary condition $\phi|_{\partial X} = (f_1, \dots, f_N)$. We define an action

$$I_X(\phi) = \frac{1}{4\pi} \iint_X d\phi \wedge *d\phi,$$

using the inner product $< , >_\Gamma$. (Locally when $d\phi = \phi_x dx + \phi_y dy$, then $*d\phi = -\phi_y dx + \phi_x dy$.) This is the generalization of the action on the cylinder we considered before.

Let $K_{X,L}$ be any positive functional which satisfies the relation

$$K_{X,L}(f_1,\ldots,f_N) = K_X \sum_{\phi_0} e^{-I_X(\phi_0)},$$

where K_X is a constant and the sum is over all harmonic maps ϕ_0 satisfying the boundary condition

$$\phi_0|_{\partial X} = (f_1,\ldots,f_N).$$

Since there is no good analogue of the Wiener measure, We can regard $K_{X,L}(f_1,\ldots,f_N)$ as a string path integral

$$\int_{\substack{\phi:X\to \mathbf{T}_L \\ \phi|_{\partial X}=(f_1,\ldots,f_N)}} e^{-I_X(\phi)}[d\phi].$$

This is because any element of the space $C^\infty{}_{X,L}(f_1,\ldots,f_N)$ can be written uniquely as a sum $\phi + \phi_0$, where $\phi : X \to \mathfrak{h}_\mathbb{R}$ is a function which satisfies the boundary condition $\phi|_{\partial X} = (0,\ldots,0)$, and ϕ_0 is a harmonic map in $C^\infty{}_{X,L}(f_1,\ldots,f_N)$, and we have the decomposition of the action

$$I_X(\phi + \phi_0) = I_X(\phi) + I_X(\phi_0).$$

The constant K_X is regarded as a string path integral

$$\int_{\substack{\phi:X\to \mathfrak{h}_\mathbb{R} \\ \phi|_{\partial X}=(0,\ldots,0)}} e^{-I_X(\phi)}[d\phi].$$

The problem is, however, there is no way of defining the constant K_X so that the Markov property (or the sewing property) is satisfied for all Riemann surfaces.

[String Path Integrals over Disks with Holes]
In Chapter II, we restrict our attention to N-holed disks P. Note that they are closed under sewing. We can define all the constants K_P in a nice way so that the Markov property is satisfied. Namely if P_3 is obtained by sewing P_1 and P_2 along a circle S, then we have the sewing of string path integrals

$$\int_{C^\infty(\mathbf{S}^1,\mathbf{T}_L)^\wedge} K_{P_1,L}(f,f_1,\ldots,f_N,g)\, K_{P_2,L}(g,h_1,\ldots,h_M)[dg]$$

$$= K_{P_3,L}(f,f_1,\ldots,f_N,h_1,\ldots,h_M).$$

where the integration is over maps g on S.

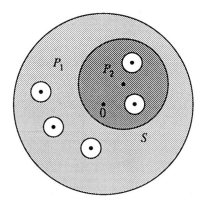

Let $\mathcal{K}$ be the Hilbert completion of $U_{L'} \subset \widetilde{\mathcal{H}}_L$. Then we have the following results.

(A) The annulus $A = \{z \in \mathbb{C} \mid |q''| \le |z| \le |q'|\}$ represents the operator

$$q^{L_0}\overline{q}^{\overline{L}_0} : \mathcal{K} \to \mathcal{K},$$

where $q = q''/q'$.

(B) The 1-holed disk $U = D_{q'} - (D_{z,\rho})^{\circ}$ represents the operator

$$q'^{-L_0}\overline{q'}^{-\overline{L}_0} \exp(zL_{-1}) \exp(\overline{z}\overline{L}_{-1})\rho^{L_0}\overline{\rho}^{\overline{L}_0} : \mathcal{K} \to \mathcal{K}.$$

(C) The 2-holed disk $P = D_{q'} - (D_{z,\rho} \cup D_{q''})^{\circ}$ represents the operator

$$q'^{-L_0}\overline{q'}^{-\overline{L}_0} Y(\rho^{L_0}\overline{\rho}^{\overline{L}_0}(\ ), z)q''^{L_0}\overline{q''}^{\overline{L}_0} : \mathcal{K} \otimes \mathcal{K} \to \mathcal{K}.$$

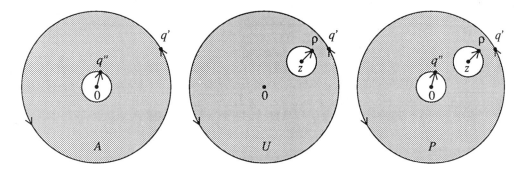

By sewing two of these 2-holed disks, we get the geometric proof of the associative law of the neutral vertex operators,

$$<v' \mid Y(Y(u, z - w)v, w) \cdot v''> = <v' \mid Y(u, z)Y(v, w) \cdot v''>,$$

when $|z| > |w|$, $0 < |z - w| < |w|$, and u, v, v', $v'' \in U_{L'}$.

[String Path Integrals over Cylinders]

In Chapter III, we adopt a more analytic approach. We define the constant K_C on cylinders using the method of zeta-regularization.

Let C be the cylinder $C_{\tau',\tau''} = \{u \in \mathbb{C} \mid \operatorname{Im}\tau' \le \operatorname{Im}u \le \operatorname{Im}\tau''\}/2\pi\mathbb{Z}$.

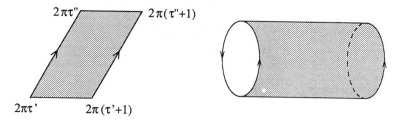

We define

$$K_C = \frac{1}{\sqrt{\det_\zeta(-A\Delta)}^\ell}.$$

as before and put

$$\widetilde{K}_{C,L}(f,g) = K_C \sum_{\phi_0} (-1)^{\phi_0} e^{-I_C(\phi_0)}.$$

where the sum is over the harmonic maps with the boundary condition $\phi_0|_{\partial C} = (f,g)$. $(-1)^{\phi_0} \in \{\pm 1\}$ is defined in a certain way. We regard it as a string path integral

$$\int_{\substack{\phi:C\to \mathbf{T}_L \\ \phi|_{\partial C}=(f,g)}} (-1)^\phi e^{-I_C(\phi)}[d\phi].$$

We have the following major properties.

(1) Let

$$d = L_0 - \frac{\ell}{24} \quad \text{and} \quad \overline{d} = \overline{L}_0 - \frac{\ell}{24}$$

be the (shifted) degree operators. $\widetilde{K}_{C,L}(f,g)$ is the integral kernel of the operator $q^d \overline{q}^{\overline{d}}$, where $q = e^{2\pi i\tau}$ and $\tau = \tau'' - \tau$. Namely, for $F \in W$, we have

$$\int_{C^\infty(\mathbf{S}^1,\mathbf{T}_L)^\wedge} \widetilde{K}_{C,L}(f,g)F(g)\, e^{-\frac{1}{2}(g_\ast,\overline{g}_\ast)}[dg] = \left(q^d\overline{q}^{\overline{d}} \cdot F\right)(f)\, e^{-\frac{1}{2}(f_\ast,\overline{f}_\ast)}.$$

Note that $d+\overline{d} = L_0 + \overline{L}_0 - \frac{\ell}{12}$ corresponds to the renormalized Hamiltonian operator H^{ren}.

We can also obtain the integral kernel

$$\widetilde{K}_{C,L}(e^{(\alpha_1,\alpha_1)},z_1)\cdots(e^{(\alpha_N,\alpha_N)},z_N)(f,g)$$

of the product of the neutral vertex operators

$$q'^{-d}\overline{q'}^{-\overline{d}}\widetilde{Y}(e^{(\alpha_1,\alpha_1)},z_1)\cdots\widetilde{Y}(e^{(\alpha_N,\alpha_N)},z_N)q''^{d}\overline{q''}^{\overline{d}} : \widetilde{\mathcal{H}}_L \to \widetilde{\mathcal{H}}_L,$$

where $\alpha_1, \ldots, \alpha_N \in L'$ and $q' = e^{2\pi i\tau'}$, $q'' = e^{2\pi i\tau''}$, which can be regarded as a string path integral

$$\int_{\substack{\phi: C \to \mathbf{T}_L \\ \phi|_{\partial C} = (f,g)}} {}_{\bullet}^{\bullet} e^{i\langle \alpha_1, \phi(z_1)\rangle} {}_{\bullet}^{\bullet} \cdots {}_{\bullet}^{\bullet} e^{i\langle \alpha_N, \phi(z_N)\rangle} {}_{\bullet}^{\bullet} (-1)^\phi e^{-I_C(\phi)} [d\phi],$$

where $z_j = e^{iu_j}$, $u_j \in C$. $(\widetilde{Y}(e^{(\alpha,\alpha)}, z) = Y(e^{(\alpha,\alpha)}, z)|z|^{\langle \alpha,\alpha\rangle}.)$

Thus we have accomplished our objective of obtaining the basic representation of the affine Lie algebra $\widehat{\mathfrak{g}}$ and the vertex operator algebra by using the string path integrals. (To obtain a string path integral realization of the product of the vertex operators $\widetilde{Y}(e^{\alpha_1}, z_1) \cdots \widetilde{Y}(e^{\alpha_N}, z_N)$, we have to take the holomorphic part of the string path integrals. This procedure will be explained elsewhere.)

(2) [Markov Property, Semigroup Property]   Let C_1 and C_2 be two cylinders and let C_3 be the cylinder obtained by sewing C_1 and C_2. Then we have

$$\int_{C^\infty(\mathbf{S}^1, \mathbf{T}_L)^\wedge} \widetilde{K}_{C_1,L}(f,g)\, \widetilde{K}_{C_2,L}(g,h)[dg] = \widetilde{K}_{C_3,L}(f,h).$$

[*String Path Integrals over Elliptic Curves*]

Let $E = E_\tau$ be the elliptic curve which is obtained by sewing the two ends of the cylinder $C_{\tau',\tau''}$, where $\tau = \tau'' - \tau$. Namely $E_\tau = \mathbb{C}/(2\pi\mathbb{Z} + 2\pi\tau\mathbb{Z})$.

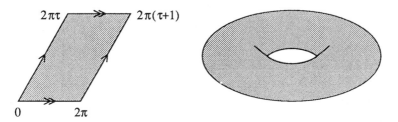

Note that any map ϕ on E with values in $\mathbf{T}_L$ can be expanded as

$$\phi(z) = \sum_{(n,m)\neq(0,0)} \phi_{n,m} e^{i[n(x - \frac{\tau_1}{\tau_2}y) - m\frac{y}{\tau_2}]} + S_{\alpha,\beta}(z) + \phi_{0,0}.$$

where $\phi_{n,m} \in \mathfrak{h}$, $\phi_{-n,-m} = \overline{\phi}_{n,m}$, $\phi_{0,0} \in \mathbf{T}_L$. We used the harmonic maps

$$S_{\alpha,\beta}(z) = \alpha \left(x - \frac{\tau_1}{\tau_2} y \right) + \beta \frac{y}{\tau_2},$$

where $z = e^{i(x+iy)}$ and $\alpha, \beta \in L$. We define $\widetilde{Z}_{E,L}$ to be equal to

$$\frac{1}{\sqrt{\det_\zeta(-A\Delta)}^\ell} \sum_{\phi_0} (-1)^{\phi_0} e^{-I_E(\phi_0)} \cdot \mathrm{vol}(\mathfrak{h}_{\mathbb{R}}/2\pi\Gamma),$$

where the sum is over all harmonic maps ϕ_0 of form $S_{\alpha,\beta}$ (without the constant term), and $(-1)^{\phi_0} = (-1)^{<\alpha,\beta>}$. We regard it as a string path integral

$$\int_{\phi:E\to\mathbf{T}_L} (-1)^\phi e^{-I_E(\phi)} [d\phi],$$

It is computed using the Kronecker's first limit formula and the Poisson summation formula.

(3) We have a string path integral realization of the trace

$$\int_{C^\infty(\mathbf{S}^1,\mathbf{T}_L)^\wedge} \widetilde{K}_{C,L}(f,f)[df] = \widetilde{Z}_{E,L} = \mathrm{Tr}_W\ q^d \bar{q}^{\bar{d}}.$$

We define a (shifted) q-character of $V_{(\omega)}$ by

$$\widetilde{ch}_q V_{(\omega)} = \mathrm{Tr}_{V_{(\omega)}}\ q^d.$$

Then the above equality (3) implies that

$$\widetilde{Z}_{E_\tau,L} = \sum_{\omega\in L'/L} |\widetilde{ch}_q V_{(\omega)}|^2.$$

Since the zeta-regularized determinant, $(-1)^{\phi_0}$ and $I_{E_\tau}(\phi_0)$ are modular invariant in τ, it is obvious that the right hand side is also modular invariant. Namely it is invariant under the transformations

$$\tau \mapsto \tau+1 \quad \text{and} \quad \tau \mapsto -\frac{1}{\tau}.$$

Note that we have to shift the degree to have the modular invariance. This clarification of the mysterious modular behavior of the (shifted) q-characters is an evidence of the power of the string path integral method.

Acknowledgement

This work was motivated by various ideas of physicists. (Most of them were communicated to the author by Professor Igor Frenkel.) Several string path integral calculations were originally done by physicists (See Hsue-Sakita-Virasoro [HSV], Polchinski [Polc]). The importance of the sewing of Riemann surfaces in general was pointed out by Segal [Seg2] in his axiomatic approach to conformal field theory and also by Witten [W1] in the bosonic string case.

I am and I shall always be very grateful to Professor Igor Frenkel. This work was not possible without his warm encouragements. His deep insight into mathematics has enlightened me. I thank Professor Nolan Wallach, Professor Bruce Driver and Professor James Lepowsky for their constructive criticism.

This paper is a slightly expanded version of my Ph. D. thesis submitted to Rutgers university in 1988.

CHAPTER I

VERTEX OPERATOR ALGEBRAS

§1. Fock Spaces

§2. Vertex Operators

§3. Representations

In this chapter, we construct the vertex operator algebras of even lattices L. In section 1, we define a Fock space V_L, which is a modification of the Fock space of string theory. Let L' be the dual lattice of L. For each element v of V_L, we define a vertex operator $Y(v,z) : V_{L'} \to V_{L'}{}^*$ in section 2. ($V_{L'}{}^*$ is the algebraic dual of $V_{L'}$.) Important properties of the vertex operators are commutative law and associative law. In section3, using these laws, we prove that the coefficients $v(n) : V_{L'} \to V_{L'}$ in $Y(v,z)$ form a Lie algebra which is called the vertex operator algebra of L. When L is a root lattice of ADE-type, the vertex operator algebra includes an affine Lie algebra $\widehat{\mathfrak{g}}$ of $\widehat{ADE}$-type as subalgebras. In fact, V_L is the basic representation of $\widehat{\mathfrak{g}}$ and $V_{L'}$ is the sum of all level $= 1$ standard representations of $\widehat{\mathfrak{g}}$.

17

§1. Fock Spaces.

In this section, we construct Fock spaces on which vertex operators act.

§1-A. Fock spaces.

DEFINITION (1.1). [Heisenberg Algebra]
Let L be an even lattice of rank ℓ. Namely, $L \cong \mathbb{Z}^\ell$ as a group, and L is equipped with a positive biadditive form $< , >$ such that $< \alpha, \alpha >$ is always even for any $\alpha \in L$. The main examples of L are the root lattices of ADE-type with $< \alpha, \alpha >= 2$ for roots α. We set $\mathfrak{h} = L \otimes_{\mathbb{Z}} \mathbb{C}$ and extend $< , >$ bilinearly to $\mathfrak{h}$. We regard $\mathfrak{h}$ as an abelian Lie algebra.
The infinite-dimensional Lie algebra

$$\widehat{\mathfrak{h}} = \mathfrak{h} \otimes \mathbb{C}[t, t^{-1}] \oplus \mathbb{C}c,$$

with the Lie brackets

$$[\alpha \otimes t^n, \beta \otimes t^m] = < \alpha, \beta > n\, \delta_{n+m,0} \cdot c,$$

$$[\alpha \otimes t^n, c] = [c, c] = 0,$$

where $\alpha, \beta \in \mathfrak{h}$ and $n, m \in \mathbb{Z}$, is called the Heisenberg Lie algebra of $\mathfrak{h}$.
We also use the notation

$$\alpha(n) = \alpha \otimes t^n, \quad \text{and} \quad \mathfrak{h}(n) = \mathfrak{h} \otimes \mathbb{C}t^n.$$

Remark. Note that the subspace $\mathfrak{h}(0)$ is central. Some people reserve the term Heisenberg Lie algebra for the Lie algebra without this subspace.

DEFINITION (1.2). [Fock Space] (See Frenkel-Kac [FK].)
Let $\widehat{\mathfrak{h}}^+ = \sum_{n=1}^{\infty} \mathfrak{h}(n)$ be the positive part of $\widehat{\mathfrak{h}}$, and $\widehat{\mathfrak{h}}^- = \sum_{n=1}^{\infty} \mathfrak{h}(-n)$ be the negative part of $\widehat{\mathfrak{h}}$. The $\mathbb{C}$-vector space

$$V_L = Sym(\widehat{\mathfrak{h}}^-) \otimes \mathbb{C}[L],$$

is called the Fock space of L. Here $Sym(\widehat{\mathfrak{h}}^-)$ is the symmetric algebra of the negative part $\widehat{\mathfrak{h}}^-$, and $\mathbb{C}[L]$ is the group algebra of the lattice L. V_L is a subspace of the larger Fock space

$$V_{L'} = Sym(\widehat{\mathfrak{h}}^-) \otimes \mathbb{C}[L'],$$

of the dual lattice L'. Namely,

$$L' = \{\beta \in \mathfrak{h}_{\mathbb{R}} \mid <\alpha,\beta> \in \mathbb{Z} \text{ for all } \alpha \in L\}.$$

We define a symmetric bilinear form $< \mid >$ on $V_{L'}$ by

$$< e^\alpha \mid e^\beta > = \delta_{\alpha,\beta} \quad \text{on } \mathbb{C}[L'],$$

$$<\alpha(-n) \mid \beta(-m)> = <\alpha,\beta> n\, \delta_{n,m} \quad \text{on } \widehat{\mathfrak{h}}^-,$$

and

$$<v_1 \cdots v_N \mid w_1 \cdots w_M> = \delta_{N,M} \cdot \sum_{\sigma \in \mathcal{S}_N} <v_1 \mid w_{\sigma(1)}> \cdots <v_N \mid w_{\sigma(N)}> \text{ on } Sym(\widehat{\mathfrak{h}}^-),$$

where $\mathcal{S}_N$ is the symmetric group of degree N. Note that $< \mid >$ is non-degenerate.

For any element ω of the dual lattice L', we define a subspace of the Fock space $V_{L'}$

$$V_{(\omega)} = Sym(\widehat{\mathfrak{h}}^-) \otimes \mathbb{C}[L + \omega].$$

Note that $V_{(0)}$ is equal to the Fock space V_L. If we fix representing elements $\lambda_0, \ldots, \lambda_k$ of L'/L (set $\lambda_0 = 0$), then the Fock space $V_{L'}$ decomposes into $(k+1)$ subspaces

$$V_{L'} \cong V_{(\lambda_0)} \oplus \cdots \oplus V_{(\lambda_k)}.$$

In the case L is a root lattice of ADE-type, L' is the corresponding weight lattice, and we can take $\{\lambda_1, \ldots, \lambda_k\}$ as the set of minuscule weights.

PROPOSITION (1.3). [Heisenberg Action]
The Heisenberg Lie algebra $\widehat{\mathfrak{h}}$ acts on the subspace $Sym(\widehat{\mathfrak{h}}^-) \otimes e^\lambda \subset V_{L'}$ canonically by

$$\alpha(-n) \cdot v \otimes e^\lambda = (\alpha(-n)v) \otimes e^\lambda,$$

$$\alpha(0) \cdot v \otimes e^\lambda = <\alpha,\lambda> v \otimes e^\lambda,$$

$$\alpha(n) \cdot v \otimes e^\lambda = \frac{\partial v}{\partial \alpha(-n)} \otimes e^\lambda,$$

for all $\alpha \in \mathfrak{h}$, $v \in Sym(\widehat{\mathfrak{h}}^-)$ and $n > 0$.

Here the operator $\dfrac{\partial}{\partial \alpha(-n)}$ is defined by

$$\frac{\partial \beta(-m)}{\partial \alpha(-n)} = <\alpha,\beta> n\, \delta_{n,m},$$

$$\frac{\partial vw}{\partial \alpha(-n)} = \frac{\partial v}{\partial \alpha(-n)} w + v \frac{\partial w}{\partial \alpha(-n)} \quad \text{(derivation)}.$$

Also c acts as Id, the identity operator.

PROPOSITION **(1.4)**.

For any operator $A : V_{L'} \to V_{L'}$, define an operator $A^\dagger : V_{L'} \to V_{L'}$ by the relation

$$< A^\dagger v' \mid v'' > = < v' \mid A v'' >,$$

for all $v', v'' \in V_{L'}$. Then for Heisenberg operators, we have

$$h(n)^\dagger = h(-n),$$

for all $h \in \mathfrak{h}$ and $n \in \mathbb{Z}$.

DEFINITION **(1.5)**.

Let $\varepsilon : L \times L \to \{\pm 1\}$ be a bilinear 2-cocycle, namely

$$\varepsilon(0, \alpha) = 1 = \varepsilon(\alpha, 0),$$
$$\varepsilon(\alpha + \beta, \gamma) = \varepsilon(\alpha, \gamma)\varepsilon(\beta, \gamma),$$
$$\varepsilon(\alpha, \beta + \gamma) = \varepsilon(\alpha, \beta)\varepsilon(\alpha, \gamma),$$

for all $\alpha, \beta, \gamma \in L$. We assume that

$$\varepsilon(\alpha, \alpha) = (-1)^{<\alpha, \alpha>/2}$$

for all $\alpha \in L$. Then we have

$$\varepsilon(\alpha, \beta)\varepsilon(\beta, \alpha) = (-1)^{<\alpha, \beta>}$$

for all $\alpha, \beta \in L$.

We define another $\mathbb{C}$-algebra structure on the $\mathbb{C}$-vector space $\mathbb{C}[L]$ by the multiplication law

$$e^\alpha \cdot e^\beta = \varepsilon(\alpha, \beta)e^{\alpha + \beta}.$$

This algebra is denoted by $\mathbb{C}[L]_\varepsilon$.

We also choose an extension of ε to $L \times L'$ with values in $\{\pm 1\}$ such that

$$\varepsilon(0, \lambda) = 1,$$
$$\varepsilon(\alpha + \beta, \lambda) = \varepsilon(\alpha, \lambda)\varepsilon(\beta, \lambda),$$
$$\varepsilon(\alpha, \beta + \lambda) = \varepsilon(\alpha, \beta)\varepsilon(\alpha, \lambda),$$

for all $\alpha, \beta \in L$ and $\lambda \in L'$. Then $\mathbb{C}$-algebra $\mathbb{C}[L]_\varepsilon$ acts on $\mathbb{C}[L']$ by

$$e^\alpha \cdot e^\lambda = \varepsilon(\alpha, \lambda)e^{\alpha + \lambda}$$

and this action preserves $\mathbb{C}[L + \omega]$ for any $\omega \in L'$. Note that we have

$$e^\alpha \cdot e^\beta = (-1)^{<\alpha, \beta>}e^\beta \cdot e^\alpha.$$

Remark : The 2-cocycle $\varepsilon : L \times L \to \{\pm 1\}$ always exists. Take a basis $\{\alpha_1, \ldots, \alpha_\ell\}$ of L, and put

$$\varepsilon(\alpha_i, \alpha_i) = (-1)^{<\alpha_i, \alpha_i>/2}, \quad \varepsilon(\alpha_i, \alpha_j) = 1 \ (i < j), \quad \varepsilon(\alpha_i, \alpha_j) = (-1)^{<\alpha_i, \alpha_j>} \ (i > j).$$

These relations define ε.

DEFINITION (1.6). [Operators L_0 and L_{-1}]
(1) We define a grading on the Fock space $V_{L'}$ with the degree defined as

$$\deg(\alpha_1(-n_1)\cdots\alpha_N(-n_N)e^{\alpha}) = n_1 + \cdots + n_N + \frac{<\alpha,\alpha>}{2}.$$

In particular, the subspace V_L is graded by $\mathbb{Z}$ and we have

$$V_L = \sum_{n=0}^{\infty} V_n, \qquad V_n = \{v \in V_L \mid \deg(v) = n\}.$$

We define a linear map

$$L_0 : V_{L'} \to V_{L'}$$

by setting

$$L_0 v = \deg(v)\, v \qquad \text{for a homogeneous } v.$$

(2) Note that $V_{L'}$ is a commutative $\mathbb{C}$-algebra. Define a linear map

$$L_{-1} : V_{L'} \to V_{L'}$$

by the equalities

$$L_{-1}e^{\alpha} = \alpha(-1)e^{\alpha},$$
$$L_{-1}\alpha(-k) = k\,\alpha(-k-1),$$

and

$$L_{-1}(vw) = L_{-1}v \cdot w + v \cdot L_{-1}w \qquad \text{(derivation)}.$$

COROLLARY (1.7).
The Fock space V_L is generated by $\mathbb{C}[L]$ as a differential $\mathbb{C}$-algebra with the derivation L_{-1}. In other words, V_L is generated by $\{L_{-1}{}^n e^{\alpha} \mid \alpha \in L,\ n \geq 0\}$ as a $\mathbb{C}$-algebra, and also by $\{L_{-1}{}^{n_1} e^{\alpha_1} \cdots L_{-1}{}^{n_N} e^{\alpha_N} \mid \alpha_1,\ldots,\alpha_N \in L,\ n_1,\ldots,n_N \geq 0\}$ as a $\mathbb{C}$-vector space.

PROPOSITION (1.8). [Character]
For any subspace W of $V_{L'}$, the q-character of W is defined by

$$ch_q W = \sum_n \dim W_n \cdot q^n, \qquad W_n = \{v \in W \mid \deg(v) = n\},$$

where q is a formal variable. Then the characters $ch_q V_{L'}$ and $ch_q V_{(\omega)}$ can be regarded as functions of q for $|q| < 1$, and we have

$$ch_q V_{L'} = \frac{\Theta_{L'}(q)}{f(q)^{\ell}},$$

$$ch_q V_{(\omega)} = \frac{\Theta_{L+\omega}(q)}{f(q)^{\ell}},$$

where

$$f(q) = \prod_{n=1}^{\infty}(1-q^n), \quad \Theta_{L'}(q) = \sum_{\alpha \in L'} q^{<\alpha,\alpha>/2}, \quad \Theta_{L+\omega}(q) = \sum_{\alpha \in L+\omega} q^{<\alpha,\alpha>/2}.$$

As we will see in the next proposition, we have to shift our grading to

$$\widetilde{ch}_q W = \sum_n \dim W_n \cdot q^{n-\frac{\ell}{24}}$$

to have a nice modular property. Then the shifted characters $\widetilde{ch}_q V_{L'}$ and $\widetilde{ch}_q V_{(\omega)}$ can be regarded as functions of τ where $q = e^{2\pi i \tau}$ for $\text{Im}\tau > 0$ and we have

$$\widetilde{ch}_q V_{L'} = \frac{\Theta_{L'}(q)}{\eta(\tau)^\ell},$$

$$\widetilde{ch}_q V_{(\omega)} = \frac{\Theta_{L+\omega} q)}{\eta(\tau)^\ell},$$

where

$$\eta(\tau) = q^{\frac{1}{24}} \prod_{n=1}^{\infty} (1 - q^n),$$

the Dedekind eta function.

The geometric meaning of these characters will be explained in sections 6 and 8. The q-character of W is also called the q-graded dimension of W and is denoted by $\dim_q W$. The shifted q-character of W is also called the shifted q-graded dimension of W and is denoted by $\widetilde{\dim}_q W$.

Proof : We will prove more general formulas in Proposition (2.20).

PROPOSITION **(1.9)**. [Modular Behavior of the Character]
The shifted q-characters have the following mysterious modular behavior in τ, where $q = e^{2\pi i \tau}$.

(1) When L is even unimodular (self-dual, i.e. $L = L'$) lattice, the absolute value

$$|\widetilde{ch}_q V_L|$$

is modular invariant in τ.

(2) For any even lattice L, the sum

$$\sum_{\omega \in L'/L} |\widetilde{ch}_q V_{(\omega)}|^2$$

is modular invariant in τ.

Here a function f of τ is called modular invariant if it is invariant under the action of the group

$$SL(2, \mathbb{Z}) = \left\{ \begin{pmatrix} a & b \\ c & d \end{pmatrix} \middle| a, b, c, d \in \mathbb{Z}, ad - bc = 1 \right\}.$$

($\begin{pmatrix} a & b \\ c & d \end{pmatrix}$ acts on τ by $\tau \mapsto \frac{a\tau + b}{c\tau + d}$.) In other words, $f(\tau + 1) = f(\tau)$ and $f(-\frac{1}{\tau}) = f(\tau)$.

Proof : See Casher-Englert-Nicolai-Taormina [CENT]. They proved the modular invariance by direct calculations. However the modular invariance suggests an underlying geometry, namely the elliptic curve E_τ. (See Example (5.23).) We will give a string path integral proof of this proposition in section 8. (See Corollary (8.4).)

§1-B. Double Fock Spaces.

DEFINITION (1.10). [Double Fock Space]

Let $\widehat{\mathfrak{h}}'$ be another copy of the Heisenberg algebra $\widehat{\mathfrak{h}}$. An element $\alpha \otimes t^n \in \widehat{\mathfrak{h}}'$ is denoted by $\alpha(n)'$. Take copies $\mathbb{C}[L]'$ and $\mathbb{C}[L']'$ of the group algebras $\mathbb{C}[L]$ and $\mathbb{C}[L']$, respectively. Elements of these copies are denoted by $(e^\alpha)'$.

We define copies of the Fock spaces

$$\overline{V_L} = Sym(\widehat{\mathfrak{h}}'^-) \otimes \mathbb{C}[L]',$$

$$\overline{V_{L'}} = Sym(\widehat{\mathfrak{h}}'^-) \otimes \mathbb{C}[L']',$$

$$\overline{V_{(\omega)}} = Sym(\widehat{\mathfrak{h}}'^-) \otimes \mathbb{C}[L+\omega]'.$$

Let $\widehat{\mathfrak{h}}^\pm = \sum_{n \neq 0} \mathfrak{h}(n) = \widehat{\mathfrak{h}}^+ \oplus \widehat{\mathfrak{h}}^-$ be the non-zero part of the Heisenberg algebra $\widehat{\mathfrak{h}}$. (See Definition (1.1).) We identify the tensor product $V_{L'} \otimes \overline{V_{L'}}$ with $Sym(\widehat{\mathfrak{h}}^\pm) \otimes \sum_{\alpha,\beta \in L'} \mathbb{C} e^{(\alpha,\beta)}$ by the correspondence

$$\alpha_1(-n_1) \cdots \alpha_N(-n_N) e^\alpha \otimes \beta_1(-m_1)' \cdots \beta_M(-m_M)'(e^\beta)'$$

$$\mapsto \alpha_1(-n_1) \cdots \alpha_N(-n_N)\beta(m_1) \cdots \beta(m_M)e^{(\alpha,\beta)}.$$

We define a $\mathbb{C}$-vector space $U_{L'}$ by

$$U_{L'} = Sym(\widehat{\mathfrak{h}}^\pm) \otimes \mathbb{C}[L'] = Sym(\widehat{\mathfrak{h}}^\pm) \otimes \sum_{\beta \in L'} \mathbb{C} e^{(\beta,\beta)} \subset V_{L'} \otimes \overline{V_{L'}}.$$

We call $U_{L'}$ the double Fock space of L'.

PROPOSITION (1.11). [Heisenberg Action]

The Heisenberg algebra $\widehat{\mathfrak{h}} \times \widehat{\mathfrak{h}}'$ acts on $Sym(\widehat{\mathfrak{h}}^\pm) \otimes e^{(r,s)} \subset V_{L'} \otimes \overline{V_{L'}}$ canonically by

$$\alpha(-n) \cdot v \otimes e^{(r,s)} = (\alpha(-n)v) \otimes e^{(r,s)},$$

$$\alpha(0) \cdot v \otimes e^{(r,s)} = <\alpha, r> v \otimes e^{(r,s)},$$

$$\alpha(n) \cdot v \otimes e^{(r,s)} = \frac{\partial v}{\partial \alpha(-n)} \otimes e^{(r,s)},$$

and

$$\alpha(-n)' \cdot v \otimes e^{(r,s)} = (\alpha(n)v) \otimes e^{(r,s)},$$

$$\alpha(0)' \cdot v \otimes e^{(r,s)} = <\alpha, s> v \otimes e^{(r,s)},$$

$$\alpha(n)' \cdot v \otimes e^{(r,s)} = \frac{\partial v}{\partial \alpha(n)} \otimes e^{(r,s)}.$$

for all $v \in Sym(\widehat{\mathfrak{h}}^\pm)$ and $n > 0$.

Here the operator $\dfrac{\partial}{\partial \alpha(n)}$ is defined by

$$\frac{\partial \beta(m)}{\partial \alpha(n)} = <\alpha, \beta>|n|\,\delta_{n,m},$$

$$\frac{\partial vw}{\partial \alpha(n)} = \frac{\partial v}{\partial \alpha(n)} w + v \frac{\partial w}{\partial \alpha(n)} \qquad \text{(derivation)},$$

for all $n, m \neq 0$.

Also c and c' acts as Id, the identity operator. (See Proposition (1.3).)

COROLLARY (1.12).

$U_{L'}$ is generated by $\mathbb{C}[L']$ as a $\mathbb{C}$-algebra with two commuting derivations L_{-1} and $\overline{L}_{-1}$. In other words, $U_{L'}$ is generated by $\{L_{-1}{}^n \overline{L}_{-1}{}^m e^{(\beta,\beta)} \mid \beta \in L',\ n, m \geq 0\}$ as a $\mathbb{C}$-algebra, and also by

$$\{L_{-1}{}^{n_1} \overline{L}_{-1}{}^{m_1} e^{(\beta_1,\beta_1)} \cdots L_{-1}{}^{n_N} \overline{L}_{-1}{}^{m_N} e^{(\beta_N,\beta_N)}$$
$$\mid \beta_1, \ldots, \beta_N \in L',\ n_1, \ldots, n_N, m_1, \ldots, m_N \geq 0\}$$

as a $\mathbb{C}$-vector space.

Here, the operator $\overline{L}_{-1}$ is defined in a same way as L_{-1}, but it is acting on $\overline{V_{L'}}$.

§2. Vertex Operators.

In this section, we define a vertex operator $Y(v, z)$ for each element $v \in V_L$. We state several key properties of the vertex operators. In particular, commutative and associative laws of vertex operators include essentially all the information about algebraic relations between coefficients of the vertex operators.

§2-A. Vertex Operators.

§2-A-1. Definition of Vertex Operators.

DEFINITION (2.1). [Boson Wick Ordering]

We regard the element $e^0 \in V_{L'}$ as a vacuum state or a ground state. Since $h(-n) \cdot e^0 \neq 0$, $(n \geq 1)$, $e^\alpha \cdot e^0 \neq 0$, we call $h(-n)$, $(n \geq 1)$ and $e^\alpha \in \mathbb{C}[L]_\varepsilon$ creation operators. Also since $h(n) \cdot e^0 = 0$, $(n \geq 0)$, we call $h(n)$, $(n \geq 0)$ annihilation operators. We define the boson Wick ordering (also called the normal ordering) as follows.

(1) For homogeneous elements $h_1, \ldots, h_n \in \widehat{\mathfrak{h}}$, we define

$$\substack{\circ \\ \circ} h_1 \cdots h_n \substack{\circ \\ \circ} = h_{\sigma(1)} \cdots h_{\sigma(i)} h_{\sigma(i+1)} \cdots h_{\sigma(n)},$$

where $\sigma \in \mathcal{S}_N$ and $h_{\sigma(1)}, \ldots, h_{\sigma(i)}$ are creation operators, $h_{\sigma(i+1)}, \ldots, h_{\sigma(n)}$ are annihilation operators. We extend this definition multilinearly. Note that the creation operators commute with each other and likewise the annihilation operators commute with each other.

$\substack{\circ \\ \circ} \ \substack{\circ \\ \circ}$ is a linear map $Sym(\widehat{\mathfrak{h}}) \to \mathcal{U}(\widehat{\mathfrak{h}})$, where $Sym(\widehat{\mathfrak{h}})$ is the symmetric algebra of the Heisenberg Lie algebra $\widehat{\mathfrak{h}}$ and $\mathcal{U}(\widehat{\mathfrak{h}})$ is the universal enveloping algebra of the Heisenberg Lie algebra $\widehat{\mathfrak{h}}$.

(2) We define

$$\substack{\circ \\ \circ} e^{\alpha_1} \cdots e^{\alpha_N} \substack{\circ \\ \circ} = e^{\alpha_1 + \cdots + \alpha_N},$$

the usual commutative product in $\mathbb{C}[L]$ and not in $\mathbb{C}[L]_\varepsilon$.

(3) Finally, we define

$$\substack{\circ \\ \circ} \alpha(0) \cdot e^\beta \substack{\circ \\ \circ} = \substack{\circ \\ \circ} e^\beta \cdot \alpha(0) \substack{\circ \\ \circ} = e^\beta \cdot \alpha(0).$$

Note that e^β commutes with $\alpha(n)$ unless $n = 0$.

Remark. In Frenkel-Kac [FK], they use a different Wick ordering. This Wick ordering used by Borcherds [Bor] is better because it yields nice formulas. (See the propositions in section 2-A-2 below.)

DEFINITION **(2.2).**
Define a formal expression

$$q_\alpha(z) = \sum_{n \neq 0} \alpha(-n) \frac{z^n}{n} + \alpha(0) \log z + \alpha,$$

for $\alpha \in \mathfrak{h}$, where z is a formal variable.

Note that we have formally

$$q_{\alpha+\beta}(z) = q_\alpha(z) + q_\beta(z),$$

$$q_{c\alpha}(z) = c\, q_\alpha(z), \quad \text{for } c \in \mathbb{C}.$$

Although $q_\alpha(z)$ does not make sense as an operator,

$$\,_\circ^\circ e^{q_\alpha(z)} \,_\circ^\circ = \exp\left(\sum_{n=1}^\infty \alpha(-n) \frac{z^n}{n}\right) e^\alpha z^{\alpha(0)} \exp\left(\sum_{n=1}^\infty \alpha(n) \frac{z^{-n}}{-n}\right), \quad \text{for } \alpha \in L,$$

and

$$\frac{d}{dz} q_\alpha(z) = \sum_{n \in \mathbb{Z}} \alpha(-n) z^{n-1}, \quad \text{for } \alpha \in \mathfrak{h},$$

are operators

$$V_{L'} \to V_{L'}{}^*$$

for $z \in \mathbb{C} - \{0\}$. (We regard e^α as an element of $\mathbb{C}[L]_\varepsilon$.) Here $V_{L'}{}^*$ is the algebraic dual of $V_{L'}$.

DEFINITION **(2.3).** [Vertex Operators] (Borcherds [Bor])
For every element $v \in V_L$ and $z \in \mathbb{C} - \{0\}$, we define an operator

$$Y(v, z) : V_{L'} \to V_{L'}{}^*,$$

by the following rules.

$$(1)\ Y(e^\alpha, z) = \,_\circ^\circ e^{q_\alpha(z)} \,_\circ^\circ, \quad \text{for } \alpha \in L,$$

$$(2)\ Y(v + w, z) = Y(v, z) + Y(w, z),$$

$$Y(cv, z) = c\, Y(v, z), \quad \text{for } c \in \mathbb{C},$$

$$(3)\ Y(v \cdot w, z) = \,_\circ^\circ Y(v, z) Y(w, z) \,_\circ^\circ,$$

$$(4)\ Y(L_{-1} v, z) = \frac{d}{dz} Y(v, z).$$

The operators $Y(v, z)$ are called vertex operators by physicists. (They use slightly different definition.) More generally, for several elements $v_1, \ldots, v_N \in V_L$ and $z_1, \ldots, z_N \in \mathbb{C} - \{0\}$, we define an operator

$$Y(v_1, \ldots, v_N; z_1, \ldots, z_N) : V_{L'} \to V_{L'}{}^*,$$

to be the Wick ordered product

$$\begin{smallmatrix}\circ\\\circ\end{smallmatrix}Y(v_1,z_1)\cdots Y(v_N,z_N)\begin{smallmatrix}\circ\\\circ\end{smallmatrix}.$$

We regard these operators as elements of $V_{L'}{}^* \otimes V_{L'}{}^*$.

PROPOSITION (2.4).

The above definition is consistent and we get

(*) $\quad Y(\alpha_1(-n_1)\cdots\alpha_N(-n_N)e^\alpha, z)$

$$= \frac{1}{(n_1-1)!\cdots(n_N-1)!}\begin{smallmatrix}\circ\\\circ\end{smallmatrix}\left(\frac{d}{dz}\right)^{n_1}q_{\alpha_1}(z)\cdots\left(\frac{d}{dz}\right)^{n_N}q_{\alpha_N}(z)\cdot e^{q_\alpha(z)}\begin{smallmatrix}\circ\\\circ\end{smallmatrix}.$$

In fact, by using the above rules, we have

$$Y(\alpha(-k-1),z) = \frac{1}{k!}Y((L_{-1})^k\cdot(L_{-1}e^\alpha\cdot e^{-\alpha}),z)$$

$$= \frac{1}{k!}\left(\frac{d}{dz}\right)^k\begin{smallmatrix}\circ\\\circ\end{smallmatrix}\frac{d}{dz}Y(e^\alpha,z)\cdot Y(e^{-\alpha},z)\begin{smallmatrix}\circ\\\circ\end{smallmatrix} = \frac{1}{k!}\left(\frac{d}{dz}\right)^{k+1}q_\alpha(z),$$

for $k \geq 0$. This implies (*). Conversely, if we assume (*) and extend it linearly, then it obviously satisfies the rules (1), (2), and (3). It is easy to see that

$$Y(L_{-1}e^\alpha, z) = \frac{d}{dz}Y(e^\alpha, z) \quad\text{and}\quad Y(L_{-1}\alpha(-n), z) = \frac{d}{dz}Y(\alpha(-n), z).$$

Also if $Y(L_{-1}v, z) = \dfrac{d}{dz}Y(v,z)$ and $Y(L_{-1}w,z) = \dfrac{d}{dz}Y(w,z)$, then

$$Y(L_{-1}(vw),z) = Y(L_{-1}v\cdot w, z) + Y(v\cdot L_{-1}w, z)$$

$$= \begin{smallmatrix}\circ\\\circ\end{smallmatrix}\frac{d}{dz}Y(v,z)\cdot Y(w,z)\begin{smallmatrix}\circ\\\circ\end{smallmatrix} + \begin{smallmatrix}\circ\\\circ\end{smallmatrix}Y(v,z)\cdot\frac{d}{dz}Y(w,z)\begin{smallmatrix}\circ\\\circ\end{smallmatrix} = \frac{d}{dz}Y(vw,z).$$

So it also satisfies (4).

Note that the the vertex operators $Y(v, z)$ preserve subspaces $V_{(\omega)}$, namely,

$$Y(v, z) : V_{(\omega)} \to V_{(\omega)}{}^*.$$

Also note that the vertex operators are parametrized by the subspace V_L and not by the space $V_{L'}$.

Remark. The essential idea of vertex operators can be found in physics literature. See Ademollo-Del Giudice-Di Vecchia- Fubini [ADDF] and Del Giudice-Di Vecchia-Fubini [DDF].

COROLLARY (2.5).

For $u = L_{-1}{}^{n_1}e^{\alpha_1}\cdots L_{-1}{}^{n_N}e^{\alpha_N}$, define a differential operator

$$D = \left(\frac{\partial}{\partial z_1}\right)^{n_1}\cdots\left(\frac{\partial}{\partial z_N}\right)^{n_N}\Bigg|_{z_1=\cdots z_N=z}.$$

Then we have

$$Y(u, z) = D\cdot Y(e^{\alpha_1},\ldots,e^{\alpha_N}; z_1,\ldots,z_N).$$

DEFINITION **(2.6).** [Coefficients of Vertex Operators]

We define

$$\widetilde{Y}(v,z) = Y(v,z)z^{\deg v},$$

for a homogeneous element $v \in V_L$. (Note that $\deg v$ is an integer.)

We define an operator

$$v(n) : V_{L'} \to V_{L'}$$

by the formula

$$\widetilde{Y}(v,z) = \sum_{n \in \mathbb{Z}} v(n)z^{-n}.$$

Remark. This operator $v(n)$ is different from the operator v_n in [Bor] defined by

$$Y(v,z) = \sum_{n \in \mathbb{Z}} v_n z^{-n-1}.$$

Examples **(2.7).**

(1) $e^0(n) = \delta_{n,0} \cdot Id$, because

$$Y(e^0, z) = Id.$$

(2) $\alpha(-1)(n) = \alpha(n)$, because

$$Y(\alpha(-1), z) = \sum_{n \in \mathbb{Z}} \alpha(n)z^{-n-1}.$$

(3) $\alpha(-2)(n) = -(n+1)\alpha(n)$, because

$$Y(\alpha(-2), z) = \sum_{n \in \mathbb{Z}} \alpha(n)(-n-1)z^{-n-2}.$$

(4) For $d = \frac{1}{2}\sum_{i=1}^{\ell} h_i(-1)^2$, where $\{h_1, \ldots, h_\ell\}$ is a real orthonormal basis of $\mathfrak{h}$, we have

$$Y(d,z) = \frac{1}{2}\sum_{i=1}^{\ell} {}^\circ_\circ\Big(\frac{d}{dz}q_{h_i}(z)\Big)^2{}^\circ_\circ.$$

Therefore,

$$d(0) = \sum_{i=1}^{\ell}\Big\{\frac{1}{2}h_i(0)^2 + \sum_{k=1}^{\infty} h_i(-k)h_i(k)\Big\} = L_0,$$

$$d(-1) = \sum_{i=1}^{\ell}\sum_{k=0}^{\infty} h_i(-k-1)h_i(k) = L_{-1}.$$

COROLLARY **(2.8).**

$$v(n) \text{ has degree } -n \text{ as an operator,}$$

namely, if $u \in V_{L'}$ has degree m, then $v(n) \cdot u$ has degree $m - n$.

COROLLARY **(2.9).**

We have the following linear relation between the coefficients of vertex operators.

$$(L_{-1}v)(n) + (L_0 v)(n) + n\,v(n) = 0.$$

$$\text{(In a different notation,} \quad (L_{-1}v)_n + n\,v_{n-1} = 0.)$$

§2-A-2. Properties of Vertex Operators.

We give several important properties of the vertex operators.

PROPOSITION (2.10).   (See Frenkel-Kac [FK] and also Frenkel-Lepowsky-Meurman [FLM2].)

Let v' and v'' be two elements of $V_{L'}$.

(1) When $\infty > |z_1| > \cdots > |z_N| > 0$, the series

$$< v' \mid Y(e^{\alpha_1}, z_1) \cdots Y(e^{\alpha_N}, z_N) \cdot v'' >$$
$$= \sum_{p_1,\ldots,p_{N-1}} < v' \mid Y(e^{\alpha_1}, z_1) \cdot p_1 > \cdots < p_{N-1} \mid Y(e^{\alpha_N}, z_N) \cdot v'' >$$

converges absolutely. Here the sum is over $N-1$ sets of real homogeneous orthonormal basis $\{p\}$ of $V_{L'}$. We have

$$< v' \mid Y(e^{\alpha_1}, z_1) \cdots Y(e^{\alpha_N}, z_N) \cdot v'' >$$
$$= \prod_{i<j} \varepsilon(\alpha_i, \alpha_j)(z_i - z_j)^{<\alpha_i, \alpha_j>} \cdot < v' \mid Y(e^{\alpha_1}, \ldots, e^{\alpha_N}; z_1, \ldots, z_N) \cdot v'' >.$$

(2) When $|z_i| > |w_j| > 0$ for all i and j, the series

$$< v' \mid Y(e^{\alpha_1}, \ldots, e^{\alpha_N}; z_1, \ldots, z_N) Y(e^{\beta_1}, \ldots, e^{\beta_M}; w_1, \ldots, w_M) \cdot v'' >$$
$$= \sum_{p} < v' \mid Y(e^{\alpha_1}, \ldots, e^{\alpha_N}; z_1, \ldots, z_N) \cdot p > < p \mid Y(e^{\beta_1}, \ldots, e^{\beta_M}; w_1, \ldots, w_M) \cdot v'' >$$

converges absolutely, and we have

$$< v' \mid Y(e^{\alpha_1}, \ldots, e^{\alpha_N}; z_1, \ldots, z_N) Y(e^{\beta_1}, \ldots, e^{\beta_M}; w_1, \ldots, w_M) \cdot v'' >$$
$$= \prod_{i,j} \varepsilon(\alpha_i, \beta_j)(z_i - w_j)^{<\alpha_i, \beta_j>}$$
$$\cdot < v' \mid Y(e^{\alpha_1}, \ldots, e^{\alpha_N}, e^{\beta_1}, \ldots, e^{\beta_M}; z_1, \ldots, z_N, w_1, \ldots, w_M) \cdot v'' >.$$

PROPOSITION (2.11).

For $v \in V_L$ and $z \in \mathbb{C} - \{0\}$, we have

$$Y(v, z) \cdot e^0 = \exp(zL_{-1}) \cdot v \in V_L{}^*.$$

In particular, we get

$$\lim_{z \to 0} Y(v, z) \cdot e^0 = v.$$

So, when v is homogeneous,

$$v(-\deg v) \cdot e^0 = v.$$

Therefore if $Y(v_1, z) = Y(v_2, z)$ as operators, then $v_1 = v_2$.

More generally, for $v_1, \ldots, v_N \in V_L$, we have

$$Y(v_1, \ldots, v_N; z_1, \ldots, z_N) \cdot e^0 = \exp(z_1 L_{-1}) v_1 \cdots \exp(z_N L_{-1}) v_N \in V_L{}^*.$$

COROLLARY (2.12).

We have

$$\mathrm{Ker} L_{-1} = \mathbb{C}e^0.$$

COROLLARY (2.13).

For $u = L_{-1}{}^{n_1} e^{\alpha_1} \cdots L_{-1}{}^{n_N} e^{\alpha_N}$, define a differential operator

$$D = \Big(\frac{\partial}{\partial z_1}\Big)^{n_1} \cdots \Big(\frac{\partial}{\partial z_N}\Big)^{n_N}\Big|_{z_1 = \cdots z_N = z}.$$

Then we have

$$u = D(Y(e^{\alpha_1}, \ldots, e^{\alpha_N}; z_1 - z, \ldots, z_N - z) \cdot e^0).$$

PROPOSITION (2.14).

For $v_1, \ldots, v_N \in V_L$ and $v', v'' \in V_{L'}$, we always have

$$<v' \mid {}^{\circ}_{\circ} Y(v_1, z_1) \cdots Y(v_N, z_N){}^{\circ}_{\circ} \cdot v''> \in \mathbb{C}\Big[z_1, \frac{1}{z_1}, \ldots, z_N, \frac{1}{z_N}\Big],$$

and immediately from the definition, we have

$$<v' \mid {}^{\circ}_{\circ} Y(v_{\sigma(1)}, z_{\sigma(1)}) \cdots Y(v_{\sigma(N)}, z_{\sigma(N)}){}^{\circ}_{\circ} \cdot v''> = <v' \mid {}^{\circ}_{\circ} Y(v_1, z_1) \cdots Y(v_N, z_N){}^{\circ}_{\circ} \cdot v''>,$$

for any permutation $\sigma \in \mathcal{S}_N$.

PROPOSITION (2.15).

Let v' and v'' be two elements of $V_{L'}$.

(1) For $u, v \in V_L$, the series

$$<v' \mid Y(Y(u, z - w)e^0 \cdot v, w) \cdot v''> = \sum_p <v' \mid Y(p, w) \cdot v''><p \mid Y(u, z - w)e^0 \cdot v>$$

converges absolutely when $|z - w| < |w|$. Here the sum is over a real homogeneous orthonormal basis $\{p\}$ of V_L. We have

$$<v' \mid Y(Y(u, z - w)e^0 \cdot v, w) \cdot v''> = <v' \mid {}^{\circ}_{\circ} Y(u, z) Y(v, w){}^{\circ}_{\circ} \cdot v''>,$$

as elements of

$$\mathbb{C}\Big[z, \frac{1}{z}, w, \frac{1}{w}\Big].$$

(2) More generally, for $u_1, \ldots, u_N, v \in V_L$, the series

$$<v' \mid Y(Y(u_1, \ldots, u_N; z_1 - w, \ldots, z_N - w)e^0 \cdot v, w) \cdot v''>$$
$$= \sum_p <v' \mid Y(p, w) \cdot v''><p \mid Y(u_1, \ldots, u_N; z_1 - w, \ldots, z_N - w)e^0 \cdot v>$$

converges absolutely when $|z_i - w| < |w|$, and we have

$$<v' \mid Y(Y(u_1, \ldots, u_N; z_1 - w, \ldots, z_N - w)e^0 \cdot v, w) \cdot v''>$$
$$= <v' \mid {}^{\circ}_{\circ} Y(u_1, \ldots, u_N; z_1, \ldots, z_N) Y(v, w){}^{\circ}_{\circ} \cdot v''>,$$

as elements of

$$\mathbb{C}\left[z_1, \frac{1}{z_1}, \ldots, z_N, \frac{1}{z_N}, w, \frac{1}{w}\right].$$

Proof : By Proposition (2.11), we have

$$<v' \mid Y(Y(u_1, \ldots, u_N; z_1 - w, \ldots, z_N - w)e^0 \cdot v, w) \cdot v''>$$
$$=<v' \mid Y(\exp((z_1 - w)L_{-1})u_1 \cdots \exp((z_N - w)L_{-1})u_N \cdot v, w) \cdot v''>$$

$$= \exp\left((z_1 - w)\frac{\partial}{\partial t_1}\right) \cdots \exp\left((z_N - w)\frac{\partial}{\partial t_N}\right)$$
$$<v' \mid {}^\circ_\circ Y(u_1, \ldots, u_N; t_1, \ldots, t_N)Y(v, w)^\circ_\circ \cdot v''>\Big|_{t_1 = \cdots = t_N = w}$$

$$=<v' \mid {}^\circ_\circ Y(u_1, \ldots, u_N; z_1, \ldots, z_N)Y(v, w)^\circ_\circ \cdot v''>.$$

PROPOSITION **(2.16)**.

(1) [Rationality of Vertex Operators]

Let $v_1, \ldots, v_N \in V_L$ and $v', v'' \in V_{L'}$. When $\infty > |z_1| > \cdots > |z_N| > 0$, the series

$$<v' \mid Y(v_1, z_1) \cdots Y(v_N, z_N) \cdot v''>$$
$$= \sum_{p_1, \ldots, p_N - 1} <v' \mid Y(v_1, z_1) \cdot p_1> \cdots <p_{N-1} \mid Y(v_N, z_N) \cdot v''>$$

converges absolutely to an element of

$$\mathbb{C}\left[z_1, \frac{1}{z_1}, \ldots, z_N, \frac{1}{z_N}, \frac{1}{z_i - z_j}\right]_{1 \leq i < j \leq N}.$$

(2) [Commutativity of Vertex Operators]

For any permutation $\sigma \in \mathcal{S}_N$, we have

$$<v' \mid Y(v_{\sigma(1)}, z_{\sigma(1)}) \cdots Y(v_{\sigma(N)}, z_{\sigma(N)}) \cdot v''>=<v' \mid Y(v_1, z_1) \cdots Y(v_N, z_N) \cdot v''>,$$

as elements of

$$\mathbb{C}\left[z_1, \frac{1}{z_1}, \ldots, z_N, \frac{1}{z_N}, \frac{1}{z_i - z_j}\right]_{1 \leq i < j \leq N}.$$

PROPOSITION **(2.17)**. [Associativity of Vertex Operators] (Borcherds [Bor], See also Frenkel-Lepowsky-Meurman [FLM2].)

For $u, v \in V_L$, and $v', v'' \in V_{L'}$, the series

$$<v' \mid Y(Y(u, z - w)v, w) \cdot v''>= \sum_p <v' \mid Y(p, w) \cdot v''><p \mid Y(u, z - w) \cdot v>,$$

converges absolutely when $0 < |z - w| < |w|$. Here the sum is over a real homogeneous orthonormal basis $\{p\}$ of V_L. The three functions

$$<v' \mid Y(Y(u, z - w)v, w) \cdot v''>,$$
$$<v' \mid Y(u, z)Y(v, w) \cdot v''>,$$
$$<v' \mid Y(v, w)Y(u, z) \cdot v''>,$$

are equal to each other as elements of

$$\mathbb{C}\left[z, \frac{1}{z}, w, \frac{1}{w}, \frac{1}{z-w}\right].$$

Proof: Because of Corollary (1.7), it is enough to prove the equation for

$$u = L_{-1}{}^{n_1} e^{\alpha_1} \cdots L_{-1}{}^{n_N} e^{\alpha_N} \quad \text{and} \quad v = L_{-1}{}^{m_1} e^{\beta_1} \cdots L_{-1}{}^{m_M} e^{\beta_M}.$$

By Proposition (2.10), we have

$$<v' \mid Y(Y(e^{\alpha_1}, \ldots, e^{\alpha_N}; z_1 - w, \ldots, z_N - w)$$
$$\cdot Y(e^{\beta_1}, \ldots, e^{\beta_M}; w_1 - w, \ldots, w_M - w)e^0, w) \cdot v''>$$
$$= \prod_{i,j} \varepsilon(\alpha_i, \beta_j)(z_i - w_j)^{<\alpha_i, \beta_j>}$$
$$\cdot <v' \mid Y(Y(e^{\alpha_1}, \ldots, e^{\alpha_N}, e^{\beta_1}, \ldots, e^{\beta_M};$$
$$z_1 - w, \ldots, z_N - w, w_1 - w, \ldots, w_M - w)e^0, w) \cdot v''>$$

for $|z_i - w| > |w_j - w|$.

When $0 < |z_i - w| < |w|$ and $0 < |w_j - w| < |w|$, it follows from Proposition (2.15) that the last formula is equal to

$$= \prod_{i,j} \varepsilon(\alpha_i, \beta_j)(z_i - w_j)^{<\alpha_i, \beta_j>}$$
$$\cdot <v' \mid Y(e^{\alpha_1}, \ldots, e^{\alpha_N}, e^{\beta_1}, \ldots, e^{\beta_M}; z_1, \ldots, z_N, w_1, \ldots, w_M) \cdot v''>.$$

Now using Proposition (2.10) again, we get

$$= <v' \mid Y(e^{\alpha_1}, \ldots, e^{\alpha_N}; z_1, \ldots, z_N) Y(e^{\beta_1}, \ldots, e^{\beta_M}; w_i, \ldots, w_M) \cdot v''>, \qquad \text{if } |z_i| > |w_j|.$$

Now apply two differential operators

$$\left(\frac{\partial}{\partial z_1}\right)^{n_1} \cdots \left(\frac{\partial}{\partial z_N}\right)^{n_N}\Big|_{z_1 = \cdots z_N = z} \quad \text{and} \quad \left(\frac{\partial}{\partial w_1}\right)^{m_1} \cdots \left(\frac{\partial}{\partial w_M}\right)^{m_M}\Big|_{w_1 = \cdots = w_M = w},$$

which correspond to u and v, then thanks to Corollaries (2.5) and (2.13), we get the proposition.

COROLLARY (2.18).

Propositions (2.15) and (2.17) imply that by comparing $Y(u, t) \cdot v$ and $Y(u, t)e^0 \cdot v$, we can find the relation between the product

$$Y(u, z)Y(v, w)$$

and the Wick ordered product

$${}^\circ_\circ Y(u, z)Y(v, w){}^\circ_\circ.$$

Here are several examples.

(1) Since we have

$$Y(e^\alpha, t) \cdot e^\beta = \varepsilon(\alpha, \beta)t^{<\alpha, \beta>} Y(e^\alpha, t)e^0 \cdot e^\beta,$$

we get

$$Y(e^\alpha, z)Y(e^\beta, w) = \varepsilon(\alpha, \beta)(z - w)^{<\alpha, \beta>} {}^\circ_\circ Y(e^\alpha, z)Y(e^\beta, w){}^\circ_\circ.$$

(2) Since we have

$$Y(\alpha(-1), t) \cdot \beta(-1) = Y(\alpha(-1), t)e^0 \cdot \beta(-1) + <\alpha, \beta> \frac{1}{t^2},$$

we get

$$Y(\alpha(-1), z)Y(\beta(-1), w) = {}^{\circ}_{\circ}Y(\alpha(-1), z)Y(\beta(-1), w){}^{\circ}_{\circ} + <\alpha, \beta> \frac{1}{(z-w)^2}.$$

(3) Since we have

$$Y(\alpha(-1), t) \cdot e^\beta = Y(\alpha(-1), t)e^0 \cdot e^\beta + <\alpha, \beta> e^\beta \frac{1}{t},$$

we get

$$Y(\alpha(-1), z)Y(e^\beta, w) = {}^{\circ}_{\circ}Y(\alpha(-1), z)Y(e^\beta, w){}^{\circ}_{\circ} + <\alpha, \beta> \frac{1}{z-w}Y(e^\beta, w).$$

COROLLARY **(2.19)**. [Vacuum Expectation Value of Product of Vertex Operators]
For any product X of vertex operators. $< e^0 | X \cdot e^0 >$ is called the vacuum expectation
value of X.
(1) We have

$$< e^0 | Y(e^{\alpha_1}, z_1) \cdots Y(e^{\alpha_N}, z_N) \cdot e^0 > = \prod_{i<j} \varepsilon(\alpha_i, \alpha_j)(z_i - z_j)^{<\alpha_i, \alpha_j>}.$$

(2) We have

$$< e^0 | Y(\alpha(-1), z)Y(\beta(-1), w) \cdot e^0 > = <\alpha, \beta> \frac{1}{(z-w)^2}.$$

PROPOSITION **(2.20)**. [Trace of Vertex Operators]
Let W be a subspace of $V_{L'}$. For a product X of vertex operators, we define the q-trace
of X by

$$\mathrm{Tr}_W X \cdot q^{L_0} = \sum_p <p \mid X \cdot q^{L_0} p>,$$

where the sum is over a real homogeneous orthonormal basis $\{p\}$ of the space W. Namely,

$$\mathrm{Tr}_W X \cdot q^{L_0} = \sum_n \mathrm{Tr}_{W_n} X \cdot q^n,$$

where $\mathrm{Tr}_{W_n} X$ is the trace of X over the space W_n. When $X = Id$, it is the q-character
of W. (See Proposition (1.8).) For example, we can compute the following q-traces. Let
$\alpha_1, \ldots, \alpha_N$ be elements of L such that $\alpha_1 + \cdots + \alpha_N = 0$. Then we have

$$\mathrm{Tr}_{V_{L'}} \widetilde{Y}(e^{\alpha_1}, z_1) \cdots \widetilde{Y}(e^{\alpha_N}, z_N) \cdot q^{L_0}$$
$$= \frac{1}{f(q)^\ell} \sum_{\gamma \in L'} (q^{\frac{1}{2}\gamma^2} z_1^{<\alpha_1, \gamma>} \cdots z_N^{<\alpha_N, \gamma>}) \cdot \prod_{i<j} \varepsilon(\alpha_i, \alpha_j)\chi_\tau(z_i, z_j)^{<\alpha_i, \alpha_j>}.$$

$$\mathrm{Tr}_{V_{(\omega)}} \widetilde{Y}(e^{\alpha_1}, z_1) \cdots \widetilde{Y}(e^{\alpha_N}, z_N) \cdot q^{L_0}$$
$$= \frac{1}{f(q)^\ell} \sum_{\gamma \in L+\omega} (q^{\frac{1}{2}\gamma^2} z_1^{<\alpha_1, \gamma>} \cdots z_N^{<\alpha_N, \gamma>}) \cdot \prod_{i<j} \varepsilon(\alpha_i, \alpha_j)\chi_\tau(z_i, z_j)^{<\alpha_i, \alpha_j>}.$$

Here

$$f(q) = \prod_{n=1}^{\infty}(1 - q^n), \quad \text{and} \quad \chi_\tau(z,w) = \frac{z - w}{\sqrt{zw}} \prod_{n=1}^{\infty} \frac{(1 - q^n\frac{z}{w})(1 - q^n\frac{w}{z})}{(1 - q^n)^2}.$$

Proof : First note that from Proposition (2.10), we have

$$\mathrm{Tr}_{V_{L'}} \, \widetilde{Y}(e^{\alpha_1}, z_1) \cdots \widetilde{Y}(e^{\alpha_N}, z_N) \cdot q^{L_0}$$
$$= \prod_{i<j} \varepsilon(\alpha_i, \alpha_j) \left(\frac{z_i - z_j}{\sqrt{z_i z_j}}\right)^{<\alpha_i, \alpha_j>} \cdot \mathrm{Tr}_{V_{L'}} \, {}^{\circ}_{\circ}Y(e^{\alpha_1}, z_1) \cdots Y(e^{\alpha_N}, z_N)^{\circ}_{\circ} \cdot q^{L_0}.$$

The Fock space $V_{L'}$ decomposes into

$$V_{L'} = \left\{ \bigotimes_{n=1}^{\infty} \bigotimes_{i=1}^{\ell} Sym(\mathbb{C}h_i(-n)) \right\} \otimes \mathbb{C}[L'],$$

where $\{h_1, \ldots, h_\ell\}$ is a real orthonormal basis of $\mathfrak{h}$, and also the operator

$${}^{\circ}_{\circ}Y(e^{\alpha_1}, z_1) \cdots Y(e^{\alpha_N}, z_N)^{\circ}_{\circ} \cdot q^{L_0}$$

decomposes into

$$\left\{ \bigotimes_{n=1}^{\infty} \bigotimes_{i=1}^{\ell} \exp\left(<\alpha_1, h_i> h_i(-n)\frac{z_1^n}{n} + \cdots + <\alpha_N, h_i> h_i(-n)\frac{z_N^n}{n} \right)\right.$$
$$\left. \cdot \exp\left(<\alpha_1, h_i> h_i(n)\frac{z_1^{-n}}{-n} + \cdots + <\alpha_N, h_i> h_i(-n)\frac{z_N^{-n}}{-n} \right) q^{L_0} \right\}$$
$$\cdot z_1^{\alpha_1(0)} \cdots z_N^{\alpha_N(0)} \cdot q^{L_0},$$

accordingly. Therefore we have

$$\mathrm{Tr} \, {}^{\circ}_{\circ}Y(e^{\alpha_1}, z_1) \cdots Y(e^{\alpha_N}, z_N)^{\circ}_{\circ} \cdot q^{L_0}$$
$$= \left\{ \prod_{n=1}^{\infty} \prod_{i=1}^{\ell} \mathrm{Tr} \, \exp\left(<\alpha_1, h_i> h_i(-n)\frac{z_1^n}{n} + \cdots + <\alpha_N, h_i> h_i(-n)\frac{z_N^n}{n} \right)\right.$$
$$\left. \cdot \exp\left(<\alpha_1, h_i> h_i(n)\frac{z_1^{-n}}{-n} + \cdots + <\alpha_N, h_i> h_i(n)\frac{z_N^{-n}}{-n} \right) q^{L_0} \right\}$$
$$\cdot \mathrm{Tr}(z_1^{\alpha_1(0)} \cdots z_N^{\alpha_N(0)} \cdot q^{L_0}).$$

Assume that $|z_1| > \cdots > |z_N| > |qz_1|$.

(1) The trace on the space $Sym(\mathbb{C}h_i(-n))$ can be computed using the orthogonal basis $\{\frac{1}{\sqrt{k!n^k}} h_i(-n)^k \mid k \geq 0\}$.

$$\mathrm{Tr} \, \exp\left(<\alpha_1, h_i> h_i(-n)\frac{z_1^n}{n} + \cdots + <\alpha_N, h_i> h_i(-n)\frac{z_N^n}{n} \right)$$
$$\cdot \exp\left(<\alpha_1, h_i> h_i(n)\frac{z_1^{-n}}{-n} + \cdots + <\alpha_N, h_i> h_i(n)\frac{z_N^{-n}}{-n} \right) q^{L_0}$$
$$= \sum_{k=0}^{\infty} \frac{1}{k!n^k} < \exp\left(<\alpha_1, h_i> h_i(n)\frac{z_1^n}{n} + \cdots + <\alpha_N, h_i> h_i(n)\frac{z_N^n}{n} \right) \cdot h_i(-n)^k$$
$$\left| \, \exp\left(<\alpha_1, h_i> h_i(n)\frac{z_1^{-n}}{-n} + \cdots + <\alpha_N, h_i> h_i(n)\frac{z_N^{-n}}{-n} \right) \cdot h_i(-n)^k > q^{nk} \right.$$

(See Proposition (1.4).)

$$= \sum_{k=0}^{\infty} \sum_{p=0}^{k} \frac{1}{(p!)^2 k! n^k}$$

$$\cdot < \frac{k!}{(k-p)!} \Big(<\alpha_1, h_i> h_i(n)\frac{z_1^n}{n} + \cdots + <\alpha_N, h_i> h_i(n)\frac{z_N^n}{n} \Big)^p \cdot h_i(-n)^{k-p}$$

$$\Big| \frac{k!}{(k-p)!} \Big(<\alpha_1, h_i> h_i(n)\frac{z_1^{-n}}{-n} + \cdots + <\alpha_N, h_i> h_i(n)\frac{z_N^{-n}}{-n} \Big)^p \cdot h_i(-n)^{k-p} > q^{nk}$$

$$= \sum_{p=0}^{\infty} \sum_{k=p}^{\infty} \frac{k!}{(p!)^2 (k-p)!} \Big(-\frac{1}{n} \sum_{I,J} <\alpha_I, h_i><\alpha_J, h_i> \Big(\frac{z_I}{z_J}\Big)^n \Big)^p q^{nk}$$

$$= \exp \Big(-\frac{1}{n} \sum_{I,J} <\alpha_I, h_i><\alpha_J, h_i> \Big(\frac{z_I}{z_J}\Big)^n \cdot \sum_{A=1}^{\infty} q^{nA} \Big) \frac{1}{1-q^n}.$$

Therefore the trace on $Sym(\widehat{\mathfrak{h}}^-)$ is

$$\prod_{n=1}^{\infty} \exp \Big(-\frac{1}{n} \sum_{I,J} <\alpha_I, \alpha_J> \Big(\frac{z_I}{z_J}\Big)^n \cdot \sum_{A=1}^{\infty} q^{nA} \Big) \frac{1}{1-q^n}$$

$$= \prod_{I<J} \Big\{ \prod_{A=1}^{\infty} \frac{(1-q^A\frac{z_I}{z_J})(1-q^A\frac{z_J}{z_I})}{(1-q^A)^2} \Big\}^{<\alpha_I,\alpha_J>} \frac{1}{f(q)^\ell}.$$

(2) The trace on $\mathbb{C}[L']$ can be computed using the orthonormal basis $\{e^\gamma \mid \gamma \in L'\}$. We get

$$\text{Tr} \, (z_1^{\alpha_1(0)} \cdots z_N^{\alpha_N(0)} \cdot q^{L_0}) = \sum_{\alpha \in L'} z_1^{<\alpha_1,\gamma>} \cdots z_N^{<\alpha_N,\gamma>} \cdot q^{\frac{1}{2}\gamma^2}.$$

(1) and (2) prove the equality for the trace on $V_{L'}$. The trace on $V_{(\omega)}$ can be computed similarly.

§2-B. Neutral Vertex Operators.

We define a variant of the vertex operators which are parametrized by the double Fock space $U_{L'}$.

DEFINITION **(2.21).** [Neutral Vertex Operators]

For any $u \in U_{L'}$, we define an operator which we call a neutral vertex operator

$$Y(u, z) : U_{L'} \to U_{L'}^*$$

where $U_{L'}^*$ is the algebraic dual of $U_{L'}$, in the following way.

(1) $Y(e^{(\beta,\beta)}, z)$

$$= \exp \left(\sum_{n=1}^{\infty} \beta(-n) \frac{z^n}{n} + \sum_{n=1}^{\infty} \beta(-n)' \frac{\overline{z}^n}{n} \right)$$

$$\cdot e^{(\beta,\beta)} |z|^{2\beta(0)} \exp \left(\sum_{n=1}^{\infty} \beta(n) \frac{z^{-n}}{-n} + \sum_{n=1}^{\infty} \beta(n)' \frac{\overline{z}^{-n}}{-n} \right), \quad \text{for } \beta \in L',$$

(2) $Y(v + w, z) = Y(v, z) + Y(w, z)$,

$\quad Y(cv, z) = c \, Y(v, z)$, for c: constant,

(3) $Y(v \cdot w, z) = {}^{\circ}_{\circ} Y(v, z) Y(w, z) {}^{\circ}_{\circ}$,

(4) $Y(L_{-1}v, z) = \dfrac{\partial}{\partial z} Y(v, z)$,

$\quad Y(\overline{L}_{-1}v, z) = \dfrac{\partial}{\partial \overline{z}} Y(v, z)$

Then for $v \otimes v' \in V_L \otimes \overline{V_L}$, we have

$$Y(v \otimes v', z) = Y(v, z) \otimes Y(v', \overline{z})'.$$

Here the operator

$$Y(v', \overline{z})' : \overline{V_{L'}}' \to (\overline{V_{L'}}')^*$$

is defined in the same way as $Y(v, z)$ but using the copies of Heisenberg operators $\alpha(n)'$ instead of $\alpha(n)$.

We also define

$$\widetilde{Y}(v, z) = Y(v, z) z^{\deg v} \overline{z}^{\deg' v}.$$

Remark. The neutral vertex operators are easier to be realized geometrically. (See section 6.) But the coefficients of the neutral vertex operators have no good algebra structures.

COROLLARY **(2.22).**

For $u = L_{-1}{}^{n_1} \overline{L}_{-1}{}^{m_1} e^{(\beta_1, \beta_1)} \cdots L_{-1}{}^{n_N} \overline{L}_{-1}{}^{m_N} e^{(\beta_N, \beta_N)}$, define a differential operator

$$D - \left(\frac{\partial}{\partial z_1} \right)^{n_1} \left(\frac{\partial}{\partial \overline{z}_1} \right)^{m_1} \cdots \left(\frac{\partial}{\partial z_N} \right)^{n_N} \left(\frac{\partial}{\partial \overline{z}_N} \right)^{m_N} \Bigg|_{z_1 = \cdots = z_N = z},$$

then we have

$$Y(u,z) = D \cdot {}^{\circ}_{\circ}Y(e^{(\beta_1,\beta_1)},z_1)\cdots Y(e^{(\beta_N,\beta_N)},z_N){}^{\circ}_{\circ}.$$

PROPOSITION **(2.23).**

We have

$$<v' \mid Y(e^{(\alpha_1,\alpha_1)},z_1)\cdots Y(e^{(\alpha_N,\alpha_N)},z_N)\cdot v''>$$

$$= \prod_{i<j}|z_i-z_j|^{2<\alpha_i,\alpha_j>} \cdot <v' \mid {}^{\circ}_{\circ}Y(e^{(\alpha_1,\alpha_1)},z_1)\cdots Y(e^{(\alpha_N,\alpha_N)},z_N){}^{\circ}_{\circ} \cdot v''>,$$

for $\infty > |z_1| > \cdots > |z_N| > 0$, and $v', v'' \in U_{L'}$.

PROPOSITION **(2.24).**

For $v \in U_{L'}$, we have

$$Y(v,z)\cdot e^{(0,0)} = \exp(zL_{-1})\exp(\overline{z}\overline{L}_{-1})\cdot v \in U_{L'}{}^{*}.$$

PROPOSITION **(2.25).**

For $u, v, v', v'' \in U_{L'}$, we have

$$<v' \mid Y(Y(u,z-w)e^{(0,0)}\cdot v,w)\cdot v''>$$

$$= <v' \mid {}^{\circ}_{\circ}Y(u,z)Y(v,w){}^{\circ}_{\circ}\cdot v''> = <v' \mid {}^{\circ}_{\circ}Y(v,w)Y(u,v){}^{\circ}_{\circ}\cdot v''>,$$

when $|z-w| < |w|$.

PROPOSITION **(2.26).** [Associative Law of Neutral Vertex Operators]

For $u, v, v', v'' \in U_{L'}$, we have the associative law of neutral vertex operators

$$<v' \mid Y(Y(u,z-w)v,w)\cdot v''> = \begin{cases} <v' \mid Y(u,z)Y(v,w)\cdot v''>, & \text{if } |z|>|w| \\ <v' \mid Y(v,w)Y(u,z)\cdot v''>, & \text{if } |z|<|w| \end{cases},$$

when $0 < |z-w| < |w|$.

§3. Representations.

In this section, we show that the operators $v(n) : V_{L'} \to V_{L'}$, where $v \in V_L$, $n \in \mathbb{Z}$ are closed under the Lie brackets. This Lie algebra of operators is called the vertex operator algebra of L.

There exist several ways to get much smaller Lie subalgebras of operators.

(i) $v \in \mathbb{C}d \oplus \mathbb{C}e^0$ gives the Virasoro algebra $\widehat{\mathcal{L}}$.

(ii) $v \in V_0 \oplus V_1$ gives the affine Lie algebra of $\mathfrak{g} = V_1$,

§3-A. Vertex Operator Algebras.

THEOREM (3.1). (Borcherds [Bor], See also Frenkel-Lepowsky-Meurman [FLM1],[FLM2].) For $u, v \in V_L$, $k, n, m \in \mathbb{Z}$, define the k-cross product

$$[u \times_k v](n, m) : V_{L'} \to V_{L'}$$

by

$$[u \times_k v](n,m) \cdot w = \sum_{i=0}^{[\deg w]-m} \binom{k}{i}(-1)^i u(n+k-i)v(m+i) \cdot w$$
$$-(-1)^k \sum_{i=0}^{[\deg w]-n} \binom{k}{i}(-1)^i v(m+k-i)u(n+i) \cdot w.$$

where w is homogeneous, and $[n]$ is the integer part of n. Then we have

$$[u \times_k v](n,m) = \sum_{i=1-\deg u}^{\deg v-k} \binom{n+\deg u-1}{i+\deg u-1}(u(i+k) \cdot v)(n+m+k),$$

when u and v are homogeneous. Here,

$$\binom{n}{i} = \frac{n(n-1)\cdots(n-i+1)}{i!} \text{ when } i > 0, \text{ and } \binom{n}{0} = 1.$$

This equality is called the Jacobi identity. Also note that

$$[u \times_0 v](n,m) = [u(n), v(m)].$$

So, in particular, the operators $v(n)$, where $v \in V_L$, $n \in \mathbb{Z}$, are closed under the Lie bracket and we have

$$[u(n), v(m)] = \sum_{i=1-\deg u}^{\deg v} \binom{n + \deg u - 1}{i + \deg u - 1} (u(i) \cdot v)(n + m),$$

when u and v are homogeneous. This Lie algebra of operators is called the vertex operator algebra of L. (This definition is slightly different from the one in [FLM2], but is essentially the same.)

Proof : Fix u, $v \in V_L$, v', $v'' \in V_{L'}$, and $w \in \mathbb{C} - \{0\}$. Let $F(z)$ be an element of $\mathbb{C}[z, \frac{1}{z}, \frac{1}{z-w}]$ which is equal to the three functions

$$<v' \mid Y(Y(u, z-w)v, w) \cdot v''>,$$
$$<v' \mid Y(u, z)Y(v, w) \cdot v''>,$$
$$<v' \mid Y(v, w)Y(u, z) \cdot v''>.$$

(See Proposition (2.17).) Then for any meromorphic function $f(z) \in \mathbb{C}[z, \frac{1}{z}, \frac{1}{z-w}]$, and a closed curve C in $\mathbb{P}^1$ which goes around the three points 0, w, and ∞ counterclockwise once each, we have

$$\frac{1}{2\pi i} \int_C F(z) f(z) \, dz = 0.$$

Take $f(z) = (z-w)^k z^{n+\deg u-1} w^{m+\deg v-1}$. Let C_ρ be the circle centered at 0 with radius ρ and assume that w is on C_ρ. Then we have

$$(*) \qquad \frac{1}{2\pi i} \int_{C_R} <v' \mid \widetilde{Y}(u, z)\widetilde{Y}(v, w) \cdot v''> (z-w)^k z^{n-1} \, dz \cdot w^{m-1}$$

$$-\frac{1}{2\pi i} \int_{C_r} <v' \mid \widetilde{Y}(v, w)\widetilde{Y}(u, z) \cdot v''> (z-w)^k z^{n-1} \, dz \cdot w^{m-1}$$

$$= \frac{1}{2\pi i} \int_{C_{w,r}} <v' \mid Y(Y(u, z-w)v, w) \cdot v''> (z-w)^k z^{n+\deg u-1} \, dz \cdot w^{m+\deg v-1},$$

where C_r and C_R are circles centered at 0 with radii r and R such that $r < \rho < R$, $C_{w,r}$ is a circle centered at w with radius r.

Integrating the left hand side of $(*)$ by w along C_ρ, we get

$$\frac{1}{(2\pi i)^2} \int_{C_\rho} \left\{ \int_{C_R} <v' \mid \widetilde{Y}(u, z)\widetilde{Y}(v, w) \cdot v''> (z-w)^k z^n \frac{dz}{z} \right\} w^m \frac{dw}{w}$$

$$-\frac{1}{(2\pi i)^2} \int_{C_\rho} \left\{ \int_{C_r} <v' \mid \widetilde{Y}(v, w)\widetilde{Y}(u, z) \cdot v''> (z-w)^k z^n \frac{dz}{z} \right\} w^m \frac{dw}{w}$$

$$= \sum_{i=0}^{\infty} \binom{k}{i} (-1)^i \frac{1}{(2\pi i)^2} \int_{C_\rho} \left\{ \int_{C_R} <v' \mid \widetilde{Y}(u,z)\widetilde{Y}(v,w) \cdot v''> z^{n+k-i} \frac{dz}{z} \right\} w^{m+i} \frac{dw}{w}$$

$$- \sum_{i=0}^{\infty} \binom{k}{i} (-1)^{k+i} \frac{1}{(2\pi i)^2} \int_{C_\rho} \left\{ \int_{C_r} <v' \mid \widetilde{Y}(v,w)\widetilde{Y}(u,z) \cdot v''> z^{n+i} \frac{dz}{z} \right\} w^{m+k-i} \frac{dw}{w}$$

$$= \sum_{i=0}^{\infty} \binom{k}{i} (-1)^i <v' \mid u(n+k-i)v(m+i) \cdot v''>$$

$$-(-1)^k \sum_{i=0}^{\infty} \binom{k}{i} (-1)^i <v' \mid v(m+k-i)u(n+i) \cdot v''>$$

$$= <v' \mid [u \times_k v](n,m) \cdot v''>$$

Integrating the right hand side of $(*)$ by w along C_ρ, we get

$$\frac{1}{(2\pi i)^2} \int_{C_\rho} \left\{ \int_{C_{w,r}} <v' \mid Y(Y(z-w,t)v,w) \cdot v''> (z-w)^k z^{n+\deg u - 1} \, dz \right\} w^{m+\deg v} \frac{dw}{w}$$

$$= \frac{1}{(2\pi i)^2} \int_{C_\rho} \left\{ \int_{C_r} <v' \mid Y(Y(u,t)v,w) \cdot v''> t^k (t+w)^{n+\deg u - 1} \, dt \right\} w^{m+\deg v} \frac{dw}{w}$$

$$= \frac{1}{(2\pi i)^2} \int_{C_\rho} \left\{ \int_{C_r} \sum_{i \in \mathbb{Z}} <v' \mid Y(u(i) \cdot v, w) \cdot v''> t^{k-i-\deg u} \right.$$
$$\left. \cdot \sum_{j \in \mathbb{Z}} \binom{n+\deg u - 1}{j} \left(\frac{t}{w}\right)^j \, dt \right\} w^{n+m+\deg u + \deg v - 1} \frac{dw}{w}$$

$$= \frac{1}{2\pi i} \int_{C_\rho} \sum_{i \in \mathbb{Z}} <v' \mid Y(u(i) \cdot v, w) \cdot v''> \binom{n+\deg u - 1}{-k+i+\deg u - 1} w^{n+m+k+\deg u(i)v} \frac{dw}{w}$$

$$= \sum_{i \in \mathbb{Z}} \binom{n+\deg u - 1}{i+\deg u - 1} <v' \mid (u(i+k) \cdot v)(n+m+k) \cdot v''>.$$

Consequently, we get

$$[u \times_k v](n,m) = \sum_{i=1-\deg u}^{\deg v - k} \binom{n+\deg u - 1}{i+\deg u - 1} (u(i+k) \cdot v)(n+m+k).$$

COROLLARY (3.2).

(1) Take $u = d$, and $k = 0$. We have $[L_{-1}, v(m)] = (L_{-1}v)(-1+m)$, and therefore,

$$[L_{-1}, Y(v,z)] = Y(L_{-1}v, z) = \frac{d}{dz} Y(v,z).$$

We have $[L_0, v(m)] = (L_{-1}v)(m) + (L_0v)(m) = -m\,v(m)$, (See Corollary (2.9).) and therefore,

$$[L_0, \widetilde{Y}(v,z)] = z\frac{d}{dz}\widetilde{Y}(v,z).$$

(2) Take $\deg u = 1$, and $k = 0$. We have $[u(0), v(m)] = (u(0) \cdot v)(m)$, and therefore,

$$[u(0), Y(v,z)] = Y(u(0) \cdot v, z).$$

COROLLARY **(3.3)**.

The operators $u(1 - \deg u)$ where u is homogeneous, are closed under the Lie bracket, and we have

$$[u(1 - \deg u), v(1 - \deg v)] = (u(1 - \deg u) \cdot v)(2 - \deg u - \deg v) = w(1 - \deg w),$$

where $w = u(1 - \deg u) \cdot v$. Note that $\deg w = \deg u + \deg v - 1$.

(In a different notation, the operators u_0 are closed under the Lie bracket, and we have

$$[u_0, v_0] = (u_0 \cdot v)_0 = w_0,$$

where $w = u_0 \cdot v$.)

§3-B. Virasoro Algebra.

THEOREM **(3.4)**. [Representation of Virasoro Algebra]

The infinite-dimensional Lie algebra

$$\widehat{\mathcal{L}} = \sum_{n\in\mathbb{Z}} \mathbb{C}L_n \oplus \mathbb{C}c'$$

with the Lie brackets

$$[L_n, L_m] = (n - m)L_{n+m} + \frac{1}{12}(n^3 - n)\,\delta_{n+m,0} \cdot c',$$

$$[L_n, c'] = [c', c'] = 0,$$

is called the Virasoro algebra. $\widehat{\mathcal{L}}$ acts on the Fock space $V_{L'}$ and also on $Sym(\widehat{\mathfrak{h}}^-) \otimes e^\lambda$ via the vertex operator

$$\sum_{n\in\mathbb{Z}} \pi(L_n) \cdot vz^{-n} = \widetilde{Y}(d,z) \cdot v, \quad \text{namely} \quad \pi(L_n) = d(n),$$

$$\text{and} \quad \pi(c') = \ell \cdot Id = \ell \cdot e^0(0),$$

where $d = \frac{1}{2}\sum_{i=1}^{\ell} h_i(-1)^2$, $\{h_1, \ldots, h_\ell\}$ is a real orthonormal basis of $\mathfrak{h}$.

Note that the operators L_0 and L_{-1} we defined in Definition (1.6) are really, $\pi(L_0)$ and $\pi(L_{-1})$.

Remark. Operators $d(n)$ are first introduced in Virasoro [V]. The computation of the Lie brackets is due to J. H. Weis (unpublished). See also Fubini-Veneziano [FV], Brower-Thorn [BT].

Proof :

(A) First we compute the following values

$$d(0) \cdot d = \left(\frac{1}{2} \sum_{i=1}^{\ell} h_i(-1)^2 \right)(0) \cdot \frac{1}{2} \sum_{i=1}^{\ell} h_i(-1)^2 = \sum_{i=1}^{\ell} h_i(-1)h_i(1) \cdot \frac{1}{2} h_i(-1)^2 = 2d,$$

$$d(1) \cdot d = \left(\frac{1}{2} \sum_{i=1}^{\ell} h_i(-1)^2 \right)(1) \cdot \frac{1}{2} \sum_{i=1}^{\ell} h_i(-1)^2 = 0,$$

$$d(2) \cdot d = \left(\frac{1}{2} \sum_{i=1}^{\ell} h_i(-1)^2 \right)(2) \cdot \frac{1}{2} \sum_{i=1}^{\ell} h_i(-1)^2 = \frac{1}{2} \sum_{i=1}^{\ell} h_i(1)^2 \cdot \frac{1}{2} \sum_{i=1}^{\ell} h_i(-1)^2 = \frac{1}{2} \ell \cdot e^0.$$

(B) Therefore, from Theorem (3.1),

$$[d(n), d(m)] = \sum_{i=-1}^{2} \binom{n+1}{i+1} (d(i) \cdot d)(n+m)$$

$$= (d(-1) \cdot d)(n+m) + (n+1)(d(0) \cdot d)(n+m)$$

$$+ \frac{(n+1)n}{2} (d(1) \cdot d)(n+m) + \frac{(n+1)n(n-1)}{6} (d(2) \cdot d)(n+m)$$

$$= (d(-1) \cdot d)(n+m) + 2(n+1)d(n+m) + \frac{(n+1)n(n-1)}{12} \ell \, \delta_{n+m,0}.$$

So ,we get

$$[d(n), d(m)] = \frac{1}{2} ([d(n), d(m)] - [d(m), d(n)])$$

$$= (n-m)d(n+m) + \frac{n^3 - n}{12} \ell \, \delta_{n+m,0}.$$

Therefore, $\pi(L_n) = d(n)$ and $\pi(c') = \ell \cdot Id$ define a representation.

§3-C. Affine Kac-Moody Lie Algebras.

THEOREM (3.5). [Representation of Affine Lie Algebras] (Frenkel-Kac [FK], Segal [Seg1]. See also Frenkel [F], Lepowsky-Primc [LP].)
Define a product $[u, v] = u(0) \cdot v$ on the degree= 1 subspace

$$V_1 = \mathfrak{h}(-1) \oplus \sum_{\alpha \in L_2} \mathbb{C}e^{\alpha}.$$

Then V_1 is a Lie algebra. (We denote it by $\mathfrak{g}$.) Here, $L_2 = \{\alpha \in L \mid < \alpha, \alpha >= 2\}$.

Let $< \, , \, >$ be the bilinear form on $\mathfrak{g}$ defined by $< u, v > e^0 = u(1) \cdot v$. (Note that $u(1) \cdot v \in V_0$.) Then $< \, , \, >$ is symmetric and invariant.

The infinite-dimensional Lie algebra

$$\widehat{\mathfrak{g}} = \mathfrak{g} \otimes \mathbb{C}[t, t^{-1}] \oplus \mathbb{C}c$$

with the Lie brackets

$$[x \otimes t^n, y \otimes t^m] = [x, y] \otimes t^{n+m} + < x, y > n\, \delta_{n+m,0} \cdot c,$$

$$[x \otimes t^n, c] = [c, c] = 0,$$

is called the affine Lie algebra of $\mathfrak{g}$.

$\widehat{\mathfrak{g}}$ acts on the Fock space $V_{L'}$ (also on $V_{(\omega)}$) via the vertex operators

$$\sum_{n \in \mathbb{Z}} \pi(u \otimes t^n) \cdot v z^{-n} = \widetilde{Y}(u, z) \cdot v, \quad \text{namely} \quad \pi(u \otimes t^n) = u(n),$$

$$\text{and} \quad \pi(c) = Id = e^0(0).$$

When L is a root lattice of ADE-type, it turns out that $\mathfrak{g}$ is the simple Lie algebra of ADE-type, and $\widehat{\mathfrak{g}}$ is the affine Lie algebra of $\widehat{ADE}$-type. Furthermore, $V_{(0)} = V_L$ is the basic representation space of $\widehat{\mathfrak{g}}$. Also, for a minuscule weight $\omega \in L'$, $V_{(\omega)}$ is the level $= 1$ standard representation space of $\widehat{\mathfrak{g}}$ corresponding to ω. Namely, $V_{(\omega)}$ is irreducible and $e^\omega \in V_{(\omega)}$ satisfies

$$\pi(x \otimes t^n) \cdot e^\omega = 0, \quad \text{for } n > 0,$$
$$\pi(e^\alpha \otimes t^0) \cdot e^\omega = 0, \quad \text{for a positive root } \alpha,$$
$$\pi(\alpha(-1) \otimes t^0) \cdot e^\omega = < \alpha, \omega > e^\omega,$$
$$\pi(c) \cdot e^\omega = e^\omega.$$

Since we have

$$\{ \text{ Minuscule weights } \} \cup \{0\} \cong L'/L,$$

it follows that $\{V_{(\omega)} \mid \omega \in L'/L\}$ is the set of all level $= 1$ standard representations of $\widehat{\mathfrak{g}}$.

Proof :

(A) First, we compute $[u, v]$ explicitly. Note that

$$u(0) \cdot v = \frac{1}{2\pi i} \int_C Y(u, z) \cdot v \, dz,$$

where C is a circle around 0.

$$[\alpha(-1), \beta(-1)] = \alpha(-1)(0) \cdot \beta(-1) = \alpha(0) \cdot \beta(-1) = 0.$$

$$[\alpha(-1), e^\beta] = \alpha(-1)(0) \cdot e^\beta = \alpha(0) \cdot e^\beta = <\alpha, \beta> e^\beta.$$

$$[e^\beta, \alpha(-1)] = e^\beta(0) \cdot \alpha(-1) = \frac{1}{2\pi i} \int_C Y(e^\beta, z) \cdot \alpha(-1) \, dz$$

$$= \frac{1}{2\pi i} \int_C \exp\left(\sum_{n=1}^{\infty} \beta(-n) \frac{z^n}{n} \right) e^\beta \left(1 - \beta(1) \frac{1}{z}\right) \cdot \alpha(-1) \, dz = -<\alpha, \beta> e^\beta.$$

$$[e^\alpha, e^\beta] = e^\alpha(0) \cdot e^\beta = \frac{1}{2\pi i} \int_C Y(e^\alpha, z) \cdot e^\beta dz$$

$$= \frac{1}{2\pi i} \int_C \exp\left(\sum_{n=1}^\infty \alpha(-n)\frac{z^n}{n}\right) \varepsilon(\alpha, \beta) e^{\alpha+\beta} z^{<\alpha,\beta>} dz$$

$$= \begin{cases} 0, & \text{if } <\alpha,\beta> \geq 0 \\ \varepsilon(\alpha,\beta)e^{\alpha+\beta}, & \text{if } <\alpha,\beta> = -1 \\ \alpha(-1)\varepsilon(\alpha,\beta)e^{\alpha+\beta} = -\alpha(-1) & \text{if } <\alpha,\beta> = -2 \Leftrightarrow \alpha+\beta = 0. \end{cases}$$

So when L is a root lattice of ADE-type, $\mathfrak{g}$ is the simple Lie algebra of ADE-type. (See Frenkel-Kac [FK], Segal [Seg1].)

Next we compute the bilinear form $< u, v >$. Note that

$$u(1) \cdot v = \frac{1}{2\pi i} \int_C Y(u, z) \cdot vz\, dz,$$

where C is a circle around 0.

$$\alpha(-1)(1) \cdot \beta(-1) = \alpha(1) \cdot \beta(-1) = <\alpha, \beta> e^0.$$

$$\alpha(-1)(1) \cdot e^\beta = \alpha(1) \cdot e^\beta = 0.$$

$$e^\beta(1) \cdot \alpha(-1) = \frac{1}{2\pi i} \int_C Y(e^\beta, z) \cdot \alpha(-1) z\, dz$$

$$= \frac{1}{2\pi i} \int_C \exp\left(\sum_{n=1}^\infty \beta(-n)\frac{z^n}{n}\right) e^\beta \left(1 - \beta(1)\frac{1}{z}\right) \cdot \alpha(-1) z\, dz = 0.$$

$$e^\alpha(1) \cdot e^\beta = \frac{1}{2\pi i} \int_C Y(e^\alpha, z) \cdot e^\beta z\, dz$$

$$= \frac{1}{2\pi i} \int_C \exp\left(\sum_{n=1}^\infty \alpha(-n)\frac{z^n}{n}\right) \varepsilon(\alpha, \beta) z^{<\alpha,\beta>} e^{\alpha+\beta} z\, dz$$

$$= \begin{cases} 0 & \text{if } <\alpha,\beta> \geq -1 \\ -e^0 & \text{if } <\alpha,\beta> = -2 \Leftrightarrow \alpha+\beta = 0. \end{cases}$$

So, we have

$$<\alpha(-1), \beta(-1)> = <\alpha, \beta>,$$
$$<\alpha(-1), e^\beta> = <e^\beta, \alpha(-1)> = 0,$$
$$<e^\alpha, e^\beta> = -\delta_{\alpha+\beta,0}.$$

Therefore, the commutation relations of the affine Lie algebra $\widehat{\mathfrak{g}}$ are given by

$$[\alpha(-1) \otimes t^n, \beta(-1) \otimes t^m] = <\alpha, \beta> n \, \delta_{n+m,0} \cdot c, \quad \text{for } \alpha, \beta \in \mathfrak{h}.$$

$$[\alpha(-1) \otimes t^n, e^\beta \otimes t^m] = <\alpha, \beta> e^\beta \otimes t^{n+m}, \quad \text{for } \alpha \in \mathfrak{h}, \beta \in L_2.$$

$$[e^\alpha \otimes t^n, e^\beta \otimes t^m]$$
$$= \begin{cases} 0 & \text{if } <\alpha, \beta> \geq 0 \\ \varepsilon(\alpha, \beta) e^{\alpha+\beta} \otimes t^{n+m} & \text{if } <\alpha, \beta> = -1 \\ -\alpha(-1) \otimes t^{n+m} - n \, \delta_{n+m,0} \cdot c & \text{if } <\alpha, \beta> = -2 \Leftrightarrow \alpha + \beta = 0, \end{cases}$$
$$\text{for } \alpha, \beta \in L_2.$$

(B) From Theorem (3.1), we get

$$[u(n), v(m)] = (u(0) \cdot v)(n + m) + n(u(1) \cdot v)(n + m)$$
$$= [u, v](n + m) + <u, v> n \, \delta_{n+m,0}.$$

Therefore $\pi(u \otimes t^n) = u(n)$ and $\pi(c) = Id$ define a representation.

Now for $\alpha \in L_2$, we get

$$\widetilde{Y}(e^\alpha, z) \cdot e^\omega = \exp\left(\sum_{n=1}^\infty \alpha(-n) \frac{z^n}{n}\right) \varepsilon(\alpha, \omega) e^{\alpha+\omega} z^{<\alpha,\omega>+1},$$

and $<\alpha, \omega> = 0$ or 1 if α is a positive root. Also for $\alpha \in \mathfrak{h}$, we get

$$\widetilde{Y}(\alpha(-1), z) \cdot e^\omega = \sum_{n=1}^\infty \alpha(-n) z^n e^\omega + <\alpha, \omega> e^\omega.$$

Therefore, the representation $V_{(\omega)}$ is exactly what we want.

PROPOSITION (3.6).

The operators $u(n)$ where $u \in \mathfrak{g}$ and the Virasoro operators $d(m)$ satisfy the commutation relation

$$[u(n), d(m)] = n \, u(n + m).$$

Proof : We have

$$\alpha(-1)(0) \cdot d = \alpha(0) \cdot d = 0, \quad \alpha(-1)(1) \cdot d = \alpha(1) \cdot d = \alpha(-1), \quad \alpha(-1)(2) \cdot d = \alpha(2) \cdot d = 0.$$

Also when $<\alpha, \alpha> = 2$, we have

$$Y(e^\alpha, z) \cdot d = \exp\left(\sum_{n=1}^\infty \alpha(-n) \frac{z^n}{n}\right) \left(e^\alpha - \frac{1}{z} \alpha(-1) e^\alpha + \frac{1}{z^2} e^0\right).$$

So we get

$$e^\alpha(0) \cdot d = 0, \quad e^\alpha(1) \cdot d = e^\alpha, \quad e^\alpha(2) \cdot d = 0.$$

Therefore for any $u \in \mathfrak{g}$, we have

$$u(0) \cdot d = 0, \quad u(1) \cdot d = u, \quad u(2) \cdot d = 0,$$

and this implies

$$[u(n), d(m)] = \sum_{i=0}^2 \binom{n}{i} (u(i) \cdot d)(n + m) = n \, u(n + m).$$

CHAPTER II

GEOMETRIC REALIZATION OF VERTEX OPERATOR ALGEBRAS

§4. Functional Realization of Fock Spaces

§5. Function Spaces on Riemann Surfaces

§6. Geometric Realization of Vertex Operators

In this chapter, we define and calculate string path integrals over N-holed disks. We realize the action of the neutral vertex operators $Y(v, z) : U_{L'} \rightarrow U_{L'}{}^*$ for $v \in U_{L'}$ using string path integrals over 2-holed disks.

In section 4, we define Gaussian measures on function spaces of $\mathbf{S}^1$, and realize the Fock spaces as pre-Hilbert spaces of functionals. We define function spaces over Riemann surfaces with boundary consisting of circles in section 5. In section 6, we show that the action of the neutral vertex operators can be realized by string path integrals over functions on a 2-holed disk. Using the sewing of two of these disks, we get a geometric proof of the associative law of the neutral vertex operators.

§4. Functional Realization of Fock Spaces.

In this section, we define Gaussian measures over function spaces on $\mathbf{S}^1$, and realize the Fock spaces as spaces of square-integrable functionals over the function spaces.

§4-A. Gaussian Measures.

First we will review basic results on Gaussian measures. See Kuo [Ku] and Gel'fand-Vilenkin [GV] for details. There are two kinds of Gaussian measures. One kind is Gaussian measure associated with an inner product on a real vector space. The other kind is Gaussian measure associated with an Hermitian form on a complex vector space.

§4-A-1. Gaussian Measures on Real Vector Spaces.

THEOREM (4.1).

Let V be an $\mathbb{R}$-vector space with an inner product $< \, , \, >$. The Hilbert completion of V with respect to $< \, , \, >$ is denoted by $\mathcal{H}$. Let $\Lambda : \mathcal{H} \to \mathcal{H}$ be a Hilbert-Schmidt operator on $\mathcal{H}$. We define a new inner product $< \, , \, >_\Lambda$ on $\mathcal{H}$ by

$$< x, y >_\Lambda = < \Lambda x, \Lambda y >.$$

Let $\mathcal{K}$ be the Hilbert completion of $\mathcal{H}$ with respect to this new inner product, and $\mathcal{K}^*$ be the dual of $\mathcal{K}$ with respect to $< \, , \, >$. We have

$$\mathcal{K}^* \subset \mathcal{H} \subset \mathcal{K}.$$

Then there exists a unique probability measure $d\mu$ on $\mathcal{K}$ such that

$$\int_{\mathcal{K}} e^{<x,y>} d\mu(x) = e^{\frac{1}{2}<y,y>},$$

for all $y \in \mathcal{K}^*$. The measure $d\mu$ is called the Gaussian measure on $\mathcal{K}$ associated with the inner product $< \, , \, >$.

PROPOSITION (4.2). [Finite Dimensional Case]

When V is finite-dimensional, we have $\mathcal{H} = V$, and we take $\Lambda = Id$. Therefore $\mathcal{K} = V$. The Gaussian measure is given by

$$d\mu(x) = e^{-\frac{1}{2}<x,x>}dx,$$

where dx is the normalized Lebesgue measure on V so that $d\mu$ is a probability.

Remark : Because of the above proposition, the integral

$$\int_{\mathcal{K}} F(x)d\mu(x),$$

is also denoted symbolically by

$$\int_{\mathcal{K}} F(x)[e^{-\frac{1}{2}<x,x>}dx],$$

even if V is infinite-dimensional.

PROPOSITION (4.3).

Let W be a finite-dimensional subspace of $\mathcal{K}$. For a function F on W, we have

$$\int_{\mathcal{K}} F(pr(x))d\mu(x) = \int_{W} F(x)\, e^{-\frac{1}{2}<x,x>}dx,$$

where pr is the projection operator onto W, and $e^{-\frac{1}{2}<x,x>}dx$ is the Gaussian measure on W.

PROPOSITION (4.4). [Quasi-Translation Invariance]

The Gaussian measure is quasi-translation invariant. Namely, for an integrable function F and an element $y \in \mathcal{K}^*$, we have

$$\int_{\mathcal{K}} F(x)d\mu(x) = \int_{\mathcal{K}} F(x+y)\, e^{-\frac{1}{2}<y,y>-<x,y>}d\mu(x).$$

PROPOSITION (4.5).

Let $\mathcal{P}(\mathcal{K}, \mathbb{R})$ be the $\mathbb{R}$-vector space of functions on $\mathcal{K}$ spanned by

$$<x,y_1>^{n_1} \cdots <x,y_N>^{n_N}\, e^{<x,y>},$$

where $y, y_1, \ldots, y_N \in \mathcal{K}^*$ and $n_1, \ldots, n_N \geq 0$. Note that the space $\mathcal{P}(\mathcal{K}, \mathbb{R})$ is also spanned by

$$D \cdot e^{<x,t_1y_1+\cdots+t_Ny_N>},$$

where

$$D = \left(\frac{\partial}{\partial t_1}\right)^{n_1} \cdots \left(\frac{\partial}{\partial t_N}\right)^{n_N}\Big|_{t_1=\cdots=t_N=0}.$$

We can compute the Gaussian integral by

$$\int_{\mathcal{K}} D \cdot e^{<x,t_1 y_1 + \cdots + t_N y_N>} d\mu(x) = D \cdot \int_{\mathcal{K}} e^{<x,t_1 y_1 + \cdots + t_N y_N>} d\mu(x).$$

DEFINITION (4.6). [Wick Ordering] (See Glimm-Jaffe [GJ].)
We can define an $\mathbb{R}$-linear map $(\,{}^\bullet_\bullet\quad{}^\bullet_\bullet\,) : \mathcal{P}(\mathcal{K}, \mathbb{R}) \to \mathcal{P}(\mathcal{K}, \mathbb{R})$ by the following two properties
(1) For $y \in \mathcal{K}^*$, we put

$$\,{}^\bullet_\bullet e^{<x,y>} \,{}^\bullet_\bullet = \frac{e^{<x,y>}}{e^{\frac{1}{2}<y,y>}}.$$

(2) For $y_1, \ldots, y_N \in \mathcal{K}^*$ and $n_1, \ldots, n_N \geq 0$, we put

$$\,{}^\bullet_\bullet D \cdot e^{<x,t_1 y_1 + \cdots + t_N y_N>} \,{}^\bullet_\bullet = D \cdot \,{}^\bullet_\bullet e^{<x,t_1 y_1 + \cdots + t_N y_N>} \,{}^\bullet_\bullet,$$

where

$$D = \left(\frac{\partial}{\partial t_1}\right)^{n_1} \cdots \left(\frac{\partial}{\partial t_N}\right)^{n_N}\bigg|_{t_1 = \cdots = t_N = 0}.$$

The function $\,{}^\bullet_\bullet F \,{}^\bullet_\bullet$ is called the Wick ordering of F with respect to the inner product $<\,,\,>$.

Examples (4.7).
(1) $\,{}^\bullet_\bullet 1 \,{}^\bullet_\bullet = 1.$
(2) $\,{}^\bullet_\bullet <x,y> \,{}^\bullet_\bullet = <x,y>.$
(3) $\,{}^\bullet_\bullet <x,y_1><x,y_2> \,{}^\bullet_\bullet = <x,y_1><x,y_2> - <y_1,y_2>.$

PROPOSITION (4.8).
(1) For $y_1, \ldots, y_N \in \mathcal{K}^*$, we have

$$\int_{\mathcal{K}} \,{}^\bullet_\bullet e^{<x,y_1>} \,{}^\bullet_\bullet \cdots \,{}^\bullet_\bullet e^{<x,y_2>} \,{}^\bullet_\bullet d\mu(x) = \prod_{i<j} e^{<y_i,y_j>}.$$

(2) For $y_1, y_2 \in \mathcal{K}^*$, we have

$$\int_{\mathcal{K}} \,{}^\bullet_\bullet <x,y_1>^n \,{}^\bullet_\bullet \,{}^\bullet_\bullet <x,y_2>^m \,{}^\bullet_\bullet d\mu(x) = n! \, \delta_{n,m} <y_1,y_2>^n.$$

PROPOSITION (4.9). (See Kuo [Ku].)
Let $A = Id + T : \mathcal{K} \to \mathcal{K}$ be a $\mathbb{R}$-linear operator such that
(1) The image of T is in $\mathcal{H}$,
(2) A is invertible as an operator $A : \mathcal{H} \to \mathcal{H}$,
(3) T is a trace class operator on $\mathcal{H}$.
Then we have

$$\int_{\mathcal{K}} F(x) d\mu(x) = \det |A| \cdot \int_{\mathcal{K}} F(Ax) \, e^{-\frac{1}{2}(<Tx,Tx>+<Tx,x>+<x,Tx>)} d\mu(x),$$

where $|A| = \sqrt{AA^*}$. Note that if $\{\lambda_1, \lambda_2, \ldots\}$ are the eigenvalues of T, then the determinant $\det |A|$ exists as the limit $\prod_{n=1}^{\infty} |1 + \lambda_n|$.

§4-A-2. Gaussian Measures on Complex Vector Spaces.

DEFINITION (4.10).

Let V be a $\mathbb{C}$-vector space with an Hermitian product $< , \overline{(\;)} >$, where $< , >$ is a symmetric bilinear form on V and $\overline{(\;)} : V \to V$ is a conjugation on V such that $< \overline{x}, \overline{y} > = \overline{< x, y >}$. The Hilbert completion of V with respect to $< , \overline{(\;)} >$ is denoted by $\mathcal{H}$. Let $\Lambda : \mathcal{H} \to \mathcal{H}$ be a Hilbert-Schmidt operator on $\mathcal{H}$. We define a new Hermitian product $< , \overline{(\;)} >_\Lambda$ on $\mathcal{H}$ by

$$< x, \overline{y} >_\Lambda = < \Lambda x, \overline{\Lambda y} >.$$

Let $\mathcal{K}$ be the Hilbert completion of $\mathcal{H}$ with respect to this new Hermitian product, and $\mathcal{K}^*$ be the dual of $\mathcal{K}$ with respect to $< , >$. We have

$$\mathcal{K}^* \subset \mathcal{H} \subset \mathcal{K}.$$

Then there exists a unique probability measure $d\mu$ on $\mathcal{K}$ such that

$$\int_{\mathcal{K}} e^{< x, y >} e^{< \overline{x}, z >} d\mu(x) = e^{< y, z >},$$

for all $y, z \in \mathcal{K}^*$. The measure $d\mu$ is called the Gaussian measure on $\mathcal{K}$ associated with the Hermitian product $< , \overline{(\;)} >$.

PROPOSITION (4.11). [Finite Dimensional Case]

When V is finite-dimensional, we have $\mathcal{H} = V$, and we take $\Lambda = Id$. Therefore $\mathcal{K} = V$. The Gaussian measure is given by

$$d\mu(x) = e^{-< x, \overline{x} >} dx\, d\overline{x},$$

where $dx\, d\overline{x}$ is the normalized Lebesgue measure on V so that $d\mu$ is a probability.

Remark : Because of the above proposition, the integral

$$\int_{\mathcal{K}} F(x) d\mu(x)$$

is also denoted symbolically by

$$\int_{\mathcal{K}} F(x) [e^{-< x, \overline{x} >} dx\, d\overline{x}]$$

even if V is infinite-dimensional.

PROPOSITION (4.12).

Let W be a finite-dimensional subspace of $\mathcal{K}$. For a function F on W, we have

$$\int_{\mathcal{K}} F(pr(x))d\mu(x) = \int_{W} F(x)\, e^{-<x,\overline{x}>} dx\, d\overline{x},$$

where pr is the projection operator onto W, and $e^{-<x,\overline{x}>} dx\, d\overline{x}$ is the Gaussian measure on W.

PROPOSITION (4.13). [Quasi-Translation Invariance]

The Gaussian measure is quasi-translation invariant. Namely, for an integrable function F and an element $y \in \mathcal{K}^*$, we have

$$\int_{\mathcal{K}} F(x)d\mu(x) = \int_{\mathcal{K}} F(x+y)\, e^{-<y,\overline{y}>-<x,\overline{y}>-<y,\overline{x}>} d\mu(x).$$

PROPOSITION (4.14).

Let $\mathcal{P}(\mathcal{K}, \mathbb{C})$ be the $\mathbb{C}$-vector space of functions on $\mathcal{K}$ spanned by

$$<x,y_1>^{n_1} \cdots <x,y_N>^{n_N}\, e^{<x,y>} \cdot <\overline{x},z_1>^{m_1} \cdots <\overline{x},z_M>^{m_M}\, e^{<\overline{x},z>},$$

where $y, y_1, \ldots, y_N, z, z_1, \ldots, z_M \in \mathcal{K}^*$ and $n_1, \ldots, n_N, m_1, \ldots, m_M \geq 0$. Note that the space $\mathcal{P}(\mathcal{K}, \mathbb{C})$ is also spanned by

$$D \cdot e^{<x,t_1 y_1 + \cdots + t_N y_N>} e^{<\overline{x}, s_1 z_1 + \cdots + s_M z_M>},$$

where

$$D = \left(\frac{\partial}{\partial t_1}\right)^{n_1} \cdots \left(\frac{\partial}{\partial t_N}\right)^{n_N} \cdot \left(\frac{\partial}{\partial s_1}\right)^{m_1} \cdots \left(\frac{\partial}{\partial s_M}\right)^{m_M} \Bigg|_{t_1 = \cdots = t_N = s_1 = \cdots = s_M = 0}.$$

We can compute the Gaussian integral by

$$\int_{\mathcal{K}} D \cdot e^{<x,t_1 y_1 + \cdots + t_N y_N>} e^{<\overline{x}, s_1 z_1 + \cdots + s_M z_M>} d\mu(x)$$

$$= D \cdot \int_{\mathcal{K}} e^{<x,t_1 y_1 + \cdots + t_N y_N>} e^{<\overline{x}, s_1 z_1 + \cdots + s_M z_M>} d\mu(x).$$

PROPOSITION (4.15).

For $y, z \in \mathcal{K}^*$, we have

$$\int_{\mathcal{K}} <x,y>^{n} <\overline{x},z>^{m}\, d\mu(x) = n!\, \delta_{n,m} <y,z>^{n}.$$

In particular, when $n \geq 1$, we have

$$\int_{\mathcal{K}} <x,y>^{n}\, d\mu(x) = 0 \qquad \text{and} \qquad \int_{\mathcal{K}} <\overline{x},z>^{n}\, d\mu(x) = 0.$$

DEFINITION (**4.16**). [Wick Ordering]

We can define a $\mathbb{C}$-linear map $({}^{\bullet}_{\bullet}\ {}^{\bullet}_{\bullet}) : \mathcal{P}(\mathcal{K}, \mathbb{C}) \to \mathcal{P}(\mathcal{K}, \mathbb{C})$ by the following two properties

(1) For $y, z \in \mathcal{K}^*$, we put

$${}^{\bullet}_{\bullet} e^{<x,y>} e^{<\overline{x},z>} {}^{\bullet}_{\bullet} = \frac{e^{<x,y>} e^{<\overline{x},z>}}{e^{<y,z>}}.$$

(2) For $y_1, \ldots, y_N, z_1, \ldots, z_M \in \mathcal{K}^*$ and $n_1, \ldots, n_N, m_1, \ldots, m_M \geq 0$, we put

$${}^{\bullet}_{\bullet} D \cdot e^{<x,t_1 y_1 + \cdots + t_N y_N>} e^{<\overline{x}, s_1 z_1 + \cdots + s_M z_M>} {}^{\bullet}_{\bullet}$$
$$= D \cdot {}^{\bullet}_{\bullet} e^{<x,t_1 y_1 + \cdots + t_N y_N>} e^{<\overline{x}, s_1 z_1 + \cdots + s_M z_M>} {}^{\bullet}_{\bullet},$$

where

$$D = \left(\frac{\partial}{\partial t_1}\right)^{n_1} \cdots \left(\frac{\partial}{\partial t_N}\right)^{n_N} \cdot \left(\frac{\partial}{\partial s_1}\right)^{m_1} \cdots \left(\frac{\partial}{\partial s_M}\right)^{m_M} \Bigg|_{t_1 = \cdots = t_N = s_1 = \cdots = s_M = 0}.$$

The function ${}^{\bullet}_{\bullet} F {}^{\bullet}_{\bullet}$ is called the Wick ordering of F with respect to the Hermitian form $< , \overline{(\ )} >$.

Examples (**4.17**).

(1) ${}^{\bullet}_{\bullet} 1 {}^{\bullet}_{\bullet} = 1$.

(2) ${}^{\bullet}_{\bullet} <x,y>^n {}^{\bullet}_{\bullet} = <x,y>^n$.

(3) ${}^{\bullet}_{\bullet} <\overline{x},z>^m {}^{\bullet}_{\bullet} = <\overline{x},z>^m$.

(4) ${}^{\bullet}_{\bullet} <x,y><\overline{x},z> {}^{\bullet}_{\bullet} = <x,y><\overline{x},z> - <y,z>$.

PROPOSITION (**4.18**).

Let $A = Id + T : \mathcal{K} \to \mathcal{K}$ be a $\mathbb{C}$-linear operator such that

(1) The image of T is in $\mathcal{H}$,

(2) A is invertible as an operator $A : \mathcal{H} \to \mathcal{H}$,

(3) T is a trace class operator on $\mathcal{H}$.

Then we have

$$\int_{\mathcal{K}} F(x) d\mu(x) = \det(AA^*) \cdot \int_{\mathcal{K}} F(Ax) \, e^{-(<Tx,\overline{Tx}>+<Tx,\overline{x}>+<x,\overline{Tx}>)} d\mu(x),$$

Note that if $\{\lambda_1, \lambda_2, \ldots\}$ are the eigenvalues of T, then the determinant $\det(AA^*)$ exists as the limit $\prod_{n=1}^{\infty}(1 + \lambda_n)(1 + \overline{\lambda}_n)$.

Remark : We can always regard a $\mathbb{C}$-vector space V as an $\mathbb{R}$-vector space $V_{(\mathbb{R})}$ by forgetting the complex structure. If we define

$$< x, y >_{\mathbb{R}} = < x, \overline{y} > + < \overline{x}, y >,$$

then $< , >_{\mathbb{R}}$ is an inner product on $V_{(\mathbb{R})}$. Since

$$\tfrac{1}{2} < x, x >_{\mathbb{R}} = < x, \overline{x} >,$$

the real and complex Gaussian integrals are compatible. Also note that

$$\mathcal{P}(\mathcal{K}_{(\mathbb{R})}, \mathbb{R}) \subset \mathcal{P}(\mathcal{K}, \mathbb{C}),$$

and the two notions of Wick ordering coincide.

§4-B. Functional Realization of Fock Spaces.

DEFINITION (4.19). [Function Spaces on $\mathbf{S}^1$]
Let L be an even lattice of rank ℓ with a positive biadditive form $< \, , \, >$ as in chapter I.
We put $\mathfrak{h}_{\mathbb{R}} = L \otimes_{\mathbb{Z}} \mathbb{R}$ and $\mathfrak{h} = L \otimes_{\mathbb{Z}} \mathbb{C}$. We extend $< \, , \, >$ to $\mathfrak{h}_{\mathbb{R}}$ and $\mathfrak{h}$. We define a new lattice

$$\Gamma = \frac{1}{\sqrt{2}} L.$$

More precisely, $\Gamma = L$ as groups, and the biadditive form $< \, , \, >_\Gamma$ of Γ is given by

$$< \alpha, \beta >_\Gamma = \frac{1}{2} < \alpha, \beta >.$$

We extend $< \, >_\Gamma$ to $\mathfrak{h}_{\mathbb{R}}$ and $\mathfrak{h}$, too. We define a torus

$$\mathbf{T}_L = \mathfrak{h}_{\mathbb{R}}/2\pi L.$$

Note that as a manifold, $\mathbf{T}_L$ is identical to $\mathfrak{h}_{\mathbb{R}}/2\pi\Gamma$. Let $\mathbf{S}^1$ be the circle parametrized as $\mathbb{R}/2\pi\mathbb{Z}$.

(1) The $\mathbb{R}$-vector space of C^∞-functions on $\mathbf{S}^1$ with values on $\mathfrak{h}_{\mathbb{R}}$ is denoted by $C^\infty(\mathbf{S}^1, \mathfrak{h}_{\mathbb{R}})$. A function f in $C^\infty(\mathbf{S}^1, \mathfrak{h}_{\mathbb{R}})$ can be expanded as

$$f(\theta) = \sum_{n \in \mathbb{Z}} f_n e^{in\theta}, \qquad f_n \in \mathfrak{h}, \quad f_{-n} = \overline{f}_n.$$

We also denote the $\mathbb{R}$-vector space of C^∞-functions on $\mathbf{S}^1$ without a constant term f_0,

$$f_*(\theta) = \sum_{n \neq 0} f_n e^{in\theta}, \qquad f_n \in \mathfrak{h}, \quad f_{-n} = \overline{f}_n,$$

by $C^\infty{}_*(\mathbf{S}^1, \mathfrak{h}_{\mathbb{R}})$. In other words,

$$C^\infty{}_*(\mathbf{S}^1, \mathfrak{h}_{\mathbb{R}}) = \left\{ f \in C^\infty(\mathbf{S}^1, \mathfrak{h}_{\mathbb{R}}) \, \middle| \, \int\limits_0^{2\pi} f(\theta)d\theta = 0 \right\}.$$

Note that there exists a natural $\mathbb{C}$-vector space structure on $C^\infty{}_*(\mathbf{S}^1, \mathfrak{h}_{\mathbb{R}})$.

(2) Let $C^\infty(\mathbf{S}^1, \mathbf{T}_L)$ be the abelian group of C^∞-maps on $\mathbf{S}^1$ with values in the torus $\mathbf{T}_L$. They are maps of the form

$$f = f_* + f_0 + \lambda\theta,$$

where $f_* \in C^\infty{}_*(\mathbf{S}^1, \mathfrak{h}_{\mathbb{R}})$, $f_0 \in \mathbf{T}_L$, and $\lambda \in L$. More specifically, if $\tilde{f} : \mathbb{R} \to \mathfrak{h}_{\mathbb{R}}$ is a lifting of f, then we have,

$$\lambda = \tilde{f}(2\pi) - \tilde{f}(0),$$

and

$$f_0 = \mathrm{pr}\left(\frac{1}{2\pi}\int_0^{2\pi}(\tilde{f}(\theta) - \lambda\theta)d\theta\right),$$

where $\mathrm{pr} : \mathfrak{h}_{\mathbb{R}} \to \mathbf{T}_L$ is the projection.

Consequently, we have an isomorphism of abelian groups

$$C^\infty(\mathbf{S}^1, \mathbf{T}_L) \cong C^\infty{}_*(\mathbf{S}^1, \mathfrak{h}_{\mathbb{R}}) \oplus \mathbf{T}_L \oplus L.$$

Let $\widetilde{C^\infty}(\mathbf{S}^1, \mathbf{T}_L)$ be an L-cover of $C^\infty(\mathbf{S}^1, \mathbf{T}_L)$,

$$\widetilde{C^\infty}(\mathbf{S}^1, \mathbf{T}_L) = C^\infty(\mathbf{S}^1, \mathfrak{h}_{\mathbb{R}}) \oplus L.$$

Elements of $\widetilde{C^\infty}(\mathbf{S}^1, \mathbf{T}_L)$ are of form

$$f = f_* + f_0 + \lambda\theta,$$

where $f_* \in C^\infty{}_*(\mathbf{S}^1, \mathfrak{h}_{\mathbb{R}})$, $f_0 \in \mathfrak{h}_{\mathbb{R}}$, and $\lambda \in L$. Note that the space $\widetilde{C^\infty}(\mathbf{S}^1, \mathbf{T}_L)$ has the following geometric meaning. Namely an element f of $\widetilde{C^\infty}(\mathbf{S}^1, \mathbf{T}_L)$ is a map f in $C^\infty(\mathbf{S}^1, \mathbf{T}_L)$ with a choice of a lifting $\tilde{f} : \mathbb{R} \to \mathfrak{h}_{\mathbb{R}}$.

DEFINITION (4.20). [Gaussian Measure on $C^\infty{}_*(\mathbf{S}^1, \mathfrak{h}_{\mathbb{R}})^\wedge$]
We define an Hermitian form $(, \overline{(\ )})$ on $C^\infty{}_*(\mathbf{S}^1, \mathfrak{h}_{\mathbb{R}})$ by the formula

$$(f_*, \overline{g}_*) = \sum_{n=1}^\infty 2n <f_n, \overline{g}_n>_\Gamma = \sum_{n=1}^\infty n <f_n, \overline{g}_n>.$$

Note that $(, \overline{(\ )})$ is also non-degenerate on $C^\infty{}_*(\mathbf{S}^1, \mathfrak{h}_{\mathbb{R}})$. We fix a positive Hilbert-Schmidt operator

$$\Lambda = \left|\frac{d}{d\theta}\right|^{-s},$$

where $s > \frac{1}{2}$ so that

$$\Lambda \cdot f_* = \sum_{n\neq 0}\frac{1}{|n|^s}f_n e^{in\theta},$$

The completion of $C^\infty{}_*(\mathbf{S}^1, \mathfrak{h}_{\mathbb{R}})$ with respect to $(, \overline{(\ )})_\Lambda$ is denoted by $C^\infty{}_*(\mathbf{S}^1, \mathfrak{h}_{\mathbb{R}})^\wedge$.

Let $d\mu(f_*)$ be the Gaussian measure on $C^\infty{}_*(\mathbf{S}^1, \mathfrak{h}_{\mathbb{R}})^\wedge$ with respect to the Hermitian form $(, \overline{(\ )})$. The Gaussian integral

$$\int_{C^\infty{}_*(\mathbf{S}^1, \mathfrak{h}_{\mathbb{R}})^\wedge} F(f_*)d\mu(f_*),$$

is also denoted symbolically by

$$\int_{C^\infty_*(\mathbf{S}^1,\mathfrak{h}_\mathbb{R})^\wedge} F(f_*)[e^{-(f_*,\overline{f}_*)}df_* d\overline{f}_*].$$

DEFINITION (4.21). [Gaussian Measures on $C^\infty(\mathbf{S}^1,\mathfrak{h}_\mathbb{R})^\wedge$ and $C^\infty(\mathbf{S}^1,\mathbf{T}_L)^\wedge$]
(1) We define a completion of $C^\infty(\mathbf{S}^1,\mathfrak{h}_\mathbb{R})$ by

$$C^\infty(\mathbf{S}^1,\mathfrak{h}_\mathbb{R})^\wedge = C^\infty_*(\mathbf{S}^1,\mathfrak{h}_\mathbb{R})^\wedge \oplus \mathfrak{h}_\mathbb{R}.$$

We use a measure

$$d\mu(f_*)\,df_0,$$

where df_0 is the Lebesgue measure on $\mathfrak{h}_\mathbb{R}$ normalized so that

$$\int_{\mathfrak{h}_\mathbb{R}} e^{-\frac{1}{2\pi}<f_0,f_0>_\mathbf{r}}df_0 = 1.$$

(2) We define a completion of $C^\infty(\mathbf{S}^1,\mathbf{T}_L)$ by

$$C^\infty(\mathbf{S}^1,\mathbf{T}_L)^\wedge = C^\infty_*(\mathbf{S}^1,\mathfrak{h}_\mathbb{R})^\wedge \oplus \mathbf{T}_L \oplus L.$$

We use a measure

$$d\mu(f) = d\mu(f_*)\,df_0\,d\lambda,$$

where df_0 is the Lebesgue measure on $\mathbf{T}_L$ induced by $\mathfrak{h}_\mathbb{R}$, and $d\lambda$ is the counting measure on L. We set

$$v_L = \int_{\mathbf{T}_L} df_0,$$

DEFINITION. (4.22). [Hilbert Space of Functionals]
(1) Let $\mathcal{H}$ be the Hilbert space of all square-integrable $\mathbb{C}$-valued functionals on the function space $C^\infty_*(\mathbf{S}^1,\mathfrak{h}_\mathbb{R})^\wedge$ with respect to the Gaussian measure $d\mu(f_*)$.

Let $\widehat{\mathfrak{h}}^\pm = \sum_{n\neq 0}\mathfrak{h}(n) = \widehat{\mathfrak{h}}^+ \oplus \widehat{\mathfrak{h}}^-$ be the degree non-zero part of the Heisenberg algebra $\widehat{\mathfrak{h}}$. (See Definition (1.1).) Then the symmetric algebra

$$Sym(\widehat{\mathfrak{h}}^\pm) = Sym(\widehat{\mathfrak{h}}^+) \otimes Sym(\widehat{\mathfrak{h}}^-)$$

can be regarded as a subspace of $\mathcal{H}$ by the equation

$$\{\alpha_1(-n_1)\cdots\alpha_N(-n_N)\beta_1(m_1)\cdots\beta_M(m_M)\}(f_*)$$
$$= {}^\bullet_\bullet in_1<\alpha_1,f_{n_1}>\cdots in_N<\alpha_N,f_{n_N}>im_1<\beta_1,\overline{f}_{m_1}>\cdots im_M<\beta_M,\overline{f}_{m_M}>{}^\bullet_\bullet,$$

where $n_1, \ldots, n_N, m_1, \ldots, m_M \geq 1$, and $\alpha_1, \ldots, \alpha_N, \beta_1, \ldots, \beta_M \in \mathfrak{h}$, $f_*(\theta) = \sum_{n \neq 0} f_n e^{in\theta}$. The Wick ordering $\vdots \ \vdots$ is with respect to the Hermitian form $(\ ,\overline{(\ )})$. (See Definition (4.16).) Note that they are polynomial functions of only finitely many variables $f_1, \ldots, f_K, \overline{f}_1, \ldots, \overline{f}_K$.

We define a bilinear form $< \ | \ >$ on $\mathcal{H}$ by

$$< F \ | \ G > = \int_{C^\infty_*(\mathbf{S}^1, \mathfrak{h}_{\mathbf{R}})^\wedge} F(f_*) G(f_*) d\mu(f_*).$$

(2) We define $\mathcal{H}_L$ to be the Hilbert space of all square-integrable $\mathbb{C}$-valued functionals on the mapping space $C^\infty(\mathbf{S}^1, \mathbf{T}_L)^\wedge$ with respect to the Gaussian measure $d\mu(f)$.

Then a vector space

$$Sym(\widehat{\mathfrak{h}}^\pm) \otimes \mathbb{C}[L'] \otimes \mathbb{C}[L].$$

can be regarded as a subspace of $\mathcal{H}_L$. $e^\beta \in \mathbb{C}[L']$ is identified with the functional

$$e^\beta(f) = e^{i<\beta, f_0>},$$

and $e^\alpha \in \mathbb{C}[L]$ is identified with the delta function

$$e^\alpha(f) = \delta_{\alpha, \lambda},$$

for $f(\theta) = \sum_{n \in \mathbb{Z}} f_n e^{in\theta} + \lambda\theta \in C^\infty(\mathbf{S}^1, \mathbf{T}_L)$.

The vector space is isomorphic to
$$Sym(\widehat{\mathfrak{h}}^\pm) \otimes \mathbb{C}[\Lambda_L],$$

where

$$\Lambda_L = \{(\beta + \tfrac{1}{2}\alpha, \beta - \tfrac{1}{2}\alpha) \ | \ \alpha \in L, \beta \in L'\}$$

is a lattice of rank 2ℓ. In other words, for

$$F = \alpha_1(-n_1) \cdots \alpha_N(-n_N)\beta_1(m_1) \cdots \beta_M(m_M)e^{(r,s)},$$

$F(f)$ is given by

$$F(f) = \delta_{r-s, \lambda} \cdot e^{i<r+s, f_0>/2} \cdot \vdots in_1 <\alpha_1, f_{n_1}> \cdots in_N <\alpha_N, f_{n_N}>$$
$$\cdot im_1 <\beta_1, \overline{f}_{m_1}> \cdots im_M <\beta_M, \overline{f}_{m_M}> \vdots.$$

We define a bilinear form $< \ | \ >$ on $\mathcal{H}_L$ by

$$< F \ | \ G > = \frac{1}{v_L} \int_{C^\infty(\mathbf{S}^1, \mathbf{T}_L)^\wedge} F(f) G(f) d\mu(f).$$

$\mathcal{H}_L$ is the Hilbert space of the closed strings moving in the torus $\mathbf{T}_L$ with respect to the metric $<\ ,\ >_\Gamma$ (not with respect to the metric $<\ ,\ >$).

(3) We define another Hilbert space $\widetilde{\mathcal{H}}_L$ to be the space of all square-integrable $\mathbb{C}$-valued functionals F on $\widetilde{C^\infty}(\mathbf{S}^1, \mathbf{T}_L)^\wedge$ which has the transformation property

$$F(f + 2\pi\mu) = (-1)^{<\lambda,\mu>} \cdot F(f),$$

for all $f(\theta) = \sum_{n \in \mathbb{Z}} f_n e^{in\theta} + \lambda\theta$ and for all $\mu \in L$. We have

$$\widetilde{\mathcal{H}}_L = F_0 \cdot \mathcal{H}_L,$$

where F_0 is the functional

$$F_0(f) = e^{i<\lambda,f_0>/2}.$$

The vector space

$$Sym(\widehat{\mathfrak{h}}^\pm) \otimes \mathbb{C}[\Omega_L],$$

where

$$\Omega_L = \{(r,s) \in L' \times L' \mid r - s \in L\}$$

is a lattice of rank 2ℓ, can be regarded as a subspace of $\widetilde{\mathcal{H}}_L$. Namely, for

$$F = \alpha_1(-n_1) \cdots \alpha_N(-n_N) \beta_1(m_1) \cdots \beta_M(m_M) e^{(r,s)},$$

$F(f)$ is given by

$$F(f) = \delta_{r-s,\lambda} \cdot e^{i<r+s,f_0>/2} \cdot {}^\bullet_\bullet in_1 <\alpha_1, f_{n_1}> \cdots in_N <\alpha_N, f_{n_N}>$$
$$\cdot im_1 <\beta_1, \overline{f}_{m_1}> \cdots im_M <\beta_M, \overline{f}_{m_M}> {}^\bullet_\bullet.$$

We define a bilinear form $< \mid >$ on $\widetilde{\mathcal{H}}_L$ by

$$< F \mid G > = \frac{1}{v_L} \int_{C^\infty(\mathbf{S}^1,\mathbf{T}_L)^\wedge} F(f)G(f) d\mu(f).$$

Note that the product FG is a functional on $C^\infty(\mathbf{S}^1, \mathbf{T}_L)^\wedge$.

Remark. Note that the double Fock space $U_{L'}$ is a subspace of $\mathcal{H}_L$ and of $\widetilde{\mathcal{H}}_L$. (See Definition (1.10).) Namely, $U_{L'}$ is isomorphic to the subspace spanned by functionals F such that

$$F(f) = \delta_{\lambda,0} \cdot e^{i<\mu,f_0>} \cdot {}^\bullet_\bullet in_1 <\alpha_1, f_{n_1}> \cdots in_N <\alpha_N, f_{n_N}>$$
$$\cdot im_1 <\beta_1, \overline{f}_{m_1}> \cdots im_M <\beta_M, \overline{f}_{m_M}> {}^\bullet_\bullet.$$

where $\mu \in L'$.

PROPOSITION (4.23).
We define

$$W = \sum_{\omega \in L'/L} V_{(\omega)} \otimes \overline{V_{(\omega)}},$$

where $\overline{V_{(\omega)}}$ is another copy of $V_{(\omega)}$. (See Definition (1.2).) Then W can be regarded as a subspace of $\widetilde{\mathcal{H}}_L$, and $< \mid >$ of W coincides with $< \mid >$ of $\widetilde{\mathcal{H}}_L$.

Proof : This is because we have

$$\mathbb{C}[\Omega_L] \cong \sum_{\omega \in L'/L} \mathbb{C}[L+\omega] \otimes \mathbb{C}[L+\omega]',$$

and therefore

$$W \cong Sym(\widehat{\mathfrak{h}}^{\pm}) \otimes \mathbb{C}[\Omega_L].$$

PROPOSITION **(4.24)**. [Functional Realization of Heisenberg Action]
Let $\widehat{\mathfrak{h}}'$ be another copy of the Heisenberg algebra $\widehat{\mathfrak{h}}$. With the above functional realization of the space W, the action of the Heisenberg algebra $\widehat{\mathfrak{h}} \times \widehat{\mathfrak{h}}'$ given in Proposition (1.11) is realized on the space of functionals $Sym(\widehat{\mathfrak{h}}^{\pm}) \otimes e^{(r,s)}$ in the following way

$$(\alpha(-n)F)(f) = \left(-i <\alpha, \frac{\partial}{\partial \overline{f}_n}> +in <\alpha, f_n>\right) F(f),$$

$$(\alpha(0)F)(f) = <\alpha, r> F(f),$$

$$(\alpha(n)F)(f) = -i <\alpha, \frac{\partial}{\partial f_n}> F(f),$$

and

$$(\alpha(-n)'F)(f) = \left(-i <\alpha, \frac{\partial}{\partial f_n}> +in <\alpha, \overline{f}_n>\right) F(f),$$

$$(\alpha(0)'F)(f) = <\alpha, s> F(f),$$

$$(\alpha(n)'F)(f) = -i <\alpha, \frac{\partial}{\partial \overline{f}_n}> F(f),$$

for all $F \in Sym(\widehat{\mathfrak{h}}^{\pm}) \otimes e^{(r,s)}$, and $n > 0$, and also

$$c = c' = Id, \text{ the identity operator.}$$

§5. Function Spaces on Riemann Surfaces.

In this section, we define function spaces on Riemann surfaces. The main point is the correspondence between the harmonic functions and the functions on the boundary. We explicitly show this in simple examples.

§5-A. General Theory.

DEFINITION (5.1). [Riemann Surface]
In this paper, the following object X is simply called a Riemann surface. X is a connected compact complex manifold of dimension $= 1$ such that the boundary ∂X consists of $N = n + m \geq 0$ circles. We fix a complex structure on X, but we do not fix a metric on X. We fix an orientation of X. We also fix a smooth parametrization of the boundary

$$\partial X \cong n\mathbf{S}^1 - m\mathbf{S}^1. \qquad \text{(The sign } \pm \text{ is according to the orientation.)}$$

The circles with the same orientation as X are called out-circles, and the circles with the opposite orientation to X are called in-circles.

When there is no boundary, we say that the Riemann surface X is closed. When there is a boundary, we say that the Riemann surface X is open.

DEFINITION (5.2). [Function Spaces on Riemann Surfaces]
Let X be a Riemann surface as in Definition (5.1).

Let $C^\infty(X, \mathfrak{h}_\mathbb{R})$ be the $\mathbb{R}$-vector space of C^∞-functions on X with values in $\mathfrak{h}_\mathbb{R}$. Also let $C^\infty(X, \mathbf{T}_L)$ be the abelian group of C^∞-maps on X with values on the torus $\mathbf{T}_L$. Also let $C^\infty(\partial X, \mathfrak{h}_\mathbb{R})$ and $C^\infty(\partial X, \mathbf{T}_L)$ be the $\mathbb{R}$-vector space of C^∞-functions on ∂X with values in $\mathfrak{h}_\mathbb{R}$ and the abelian group of C^∞-maps on ∂X with values in $\mathbf{T}_L$, respectively.

For functions on the circle $f_1, \ldots, f_N \in C^\infty(\mathbf{S}^1, \mathfrak{h}_\mathbb{R})$, the set of C^∞-functions ϕ on X with values in $\mathfrak{h}_\mathbb{R}$ with the boundary condition

$$\phi = (f_1, \ldots, f_N) \qquad \text{on } \partial X,$$

is denoted by $C^\infty{}_X(f_1, \ldots, f_N)$. Note that the set $C^\infty{}_X(f_1, \ldots, f_N)$ is not an $\mathbb{R}$-vector space unless we have $f_1 = \cdots = f_N = 0$.

For maps on the circle $f_1, \ldots, f_N \in C^\infty(\mathbf{S}^1, \mathbf{T}_L)$, the set of C^∞-maps ϕ on X with values in $\mathbf{T}_L$ with the boundary condition

$$\phi = (f_1, \ldots, f_N) \qquad \text{on } \partial X,$$

is denoted by $C^\infty{}_{X,L}(f_1,\ldots,f_N)$. Note that the set $C^\infty{}_{X,L}(f_1,\ldots,f_N)$ is not an abelian group unless we have $f_1 = \cdots = f_N = 0$.

DEFINITION (5.3). [Dirichlet Inner Product]
Let g be a Riemannian metric on X which is compatible with the complex structure on X. Namely, g can be expressed locally as $g = e^\rho dz d\bar{z} = e^\rho(dx^2 + dy^2)$ using a local coordinate $z = x + iy$, and a function ρ.

Let ω_g be the area form associated to g, namely ω_g is locally $\omega_g = \sqrt{g}\,dxdy$, where $\sqrt{g}$ means $\sqrt{\det g}$. Also let $Area_g(X)$ be the area of X with respect to the metric g, namely

$$Area_g(X) = \iint_X \omega_g.$$

Then we can define a bilinear form $<\,,\,>$ on $C^\infty(X, \mathfrak{h}_\mathbb{R})$ by

$$<\phi, \psi> = \frac{1}{Area_g(X)} \iint_X <\phi, \psi>_\Gamma \omega_g,$$

which does depend on the metric g.

We define a vector space $C^\infty{}_*(X, \mathfrak{h}_\mathbb{R})$ by

$$C^\infty{}_*(X, \mathfrak{h}_\mathbb{R}) = \left\{ \phi \in C^\infty(X, \mathfrak{h}_\mathbb{R}) \,\middle|\, \iint_X \phi \omega_g = 0 \right\}.$$

We also define so-called Dirichlet inner product $\ll\,,\,\gg$ on $C^\infty(X, \mathfrak{h}_\mathbb{R})$ and on $C^\infty(X, \mathbf{T}_L)$ by

$$\ll \phi, \psi \gg = \iint_X \sum_{\alpha,\beta} g^{\alpha\beta} <\partial_\alpha \phi, \partial_\beta \psi>_\Gamma \omega_g = \iint_X d\phi \wedge *d\phi,$$

where $(g^{\alpha\beta})$ is the inverse matrix of $(g_{\alpha\beta})$.

Note that if there exists a global coordinate $z = x + iy$, then

$$\ll \phi, \psi \gg = \iint_X (<\phi_x, \psi_x>_\Gamma + <\phi_y, \psi_y>_\Gamma)\,dxdy.$$

Also note that $\ll \phi, \psi \gg$ depends only on the complex structure on X and does not depend on the metric g, namely, $\ll\,,\,\gg$ is conformally invariant.

We put

$$I_X(\phi) = \frac{1}{4\pi} \ll \phi, \phi \gg,$$

which is called the action integral.

DEFINITION (5.4). [Harmonic Functions, Harmonic Maps]
The Laplacian operator is defined by

$$\Delta_g = \frac{1}{\sqrt{g}} \sum_{\alpha,\beta} \partial_\alpha(\sqrt{g} g^{\alpha\beta} \partial_\beta).$$

When $g = e^\rho(dx^2 + dy^2)$, we have

$$\Delta_g = e^{-\rho}\left(\left(\frac{\partial}{\partial x}\right)^2 + \left(\frac{\partial}{\partial y}\right)^2\right).$$

Note that Laplacian depends on the metric g.

A function $\phi\colon X \to \mathfrak{h}_{\mathbb{R}}$ or a map $\phi\colon X \to \mathbf{T}_L$ is called harmonic if it satisfies

$$\Delta_g \phi = 0.$$

Let $\mathrm{Harm}(X, \mathfrak{h}_{\mathbb{R}})$ be the $\mathbb{R}$-vector space of harmonic functions on X with values in $\mathfrak{h}_{\mathbb{R}}$, and let $\mathrm{Harm}(X, \mathbf{T}_L)$ be the abelian group of harmonic maps on X with values in $\mathbf{T}_L$. We also set

$$\mathrm{Harm}_{X,L}(f_1, \ldots, f_N) = \mathrm{Harm}(X, \mathbf{T}_L) \cap C^\infty{}_{X,L}(f_1, \ldots, f_N),$$

namely the set of harmonic maps in $C^\infty{}_{X,L}(f_1, \ldots, f_N)$.

Note that $\mathrm{Harm}(X, \mathfrak{h}_{\mathbb{R}})$ and $\mathrm{Harm}(X, \mathbf{T}_L)$ are independent of the metric g, and depend only on the conformal structure of X.

§5-A-1. Functions on Riemann Surface.

Now we state several general facts about the functions on X with values in the $\mathbb{R}$-vector space $\mathfrak{h}_{\mathbb{R}}$.

PROPOSITION (5.5).
When X is closed, the only harmonic functions on X are constant functions, namely we have

$$\mathrm{Harm}(X, \mathfrak{h}_{\mathbb{R}}) \cong \mathfrak{h}_{\mathbb{R}}.$$

PROPOSITION (5.6). [Solvability of Dirichlet Problem]
Let X be an open Riemann surface. For any functions on the boundary $f_1, \ldots, f_N \in C^\infty(\mathbf{S}^1, \mathfrak{h}_{\mathbb{R}})$, there is a unique harmonic function ϕ_0 on X such that

$$\phi_0 = (f_1, \ldots, f_N), \quad \text{on } \partial X.$$

Thus we have a one to one correspondence between the harmonic functions on X and the functions on the boundary ∂X. Namely we have the $\mathbb{R}$-linear isomorphism

$$\mathrm{Harm}(X, \mathfrak{h}_{\mathbb{R}}) \cong C^\infty(\partial X, \mathfrak{h}_{\mathbb{R}}).$$

We denote the above unique harmonic function ϕ_0 by $\phi_X(f_1, \ldots, f_N)$.

Proof : See Hörmander [Hö] p.264.

PROPOSITION (5.7). [Green Orthogonal Decomposition]
When a Riemann surface X is open, by the Green's theorem and the solvability of Dirichlet problem, we have an orthogonal decomposition with respect to the Dirichlet inner product $\ll , \gg$

$$C^\infty(X, \mathfrak{h}_{\mathbb{R}}) \cong C^\infty{}_X(0, \ldots, 0) \oplus \operatorname{Harm}(X, \mathfrak{h}_{\mathbb{R}}).$$

Proof : Green's theorem states that for all functions $\phi, \psi \in C^\infty(X, \mathfrak{h}_{\mathbb{R}})$, we have

$$\ll \phi, \psi \gg = \int_{\partial X} \phi * d\psi - \iint_X <\phi, \Delta_g \psi>_\Gamma \omega_g,$$

where $*d\psi = -\psi_y dx + \psi_x dy$, locally.

   Therefore if $\phi = 0$ on ∂X and $\Delta_g \psi = 0$ on X, then ϕ and ψ are orthogonal to each other. Also for $\phi \in C^\infty(X, \mathfrak{h}_{\mathbb{R}})$, the decomposition into non-harmonic and harmonic part is given by

$$\phi \mapsto (\phi - \phi_0, \phi_0),$$

where ϕ_0 is the unique harmonic function such that $\phi = \phi_0$ on ∂X.

COROLLARY (5.8).
Let X be an open Riemann surface. For functions on the boundary $f_1, \ldots, f_N \in C^\infty(\mathbf{S}^1, \mathfrak{h}_{\mathbb{R}})$, the set $C^\infty{}_X(f_1, \ldots, f_N)$ has a structure of affine space with the map

$$C^\infty{}_X(f_1, \ldots, f_N) \cong C^\infty{}_X(0, \ldots, 0) \; : \; \phi + \phi_0 \leftrightarrow \phi,$$

where $\phi_0 = \phi_X(f_1, \ldots, f_N)$.

§5-A-2. Maps on Riemann Surface.

Now we state several general facts about the maps on X with values in the torus $\mathbf{T}_L$.

DEFINITION (5.9). [Homotopy Type of Maps]
Two maps $\phi, \psi \in C^\infty(X, \mathbf{T}_L)$ are said to be homotopic if there exists a continuous map $F : I \times X \to \mathbf{T}_L$, where $I = [0, 1]$, the unit interval, such that

$$F(0, -) = \phi \quad \text{and} \quad F(1, -) = \psi.$$

F is called a homotopy connecting ϕ and ψ. We write $\phi \sim \psi$ when ϕ and ψ are homotopic.
   A homotopy class of maps from X to $\mathbf{T}_L$ is an equivalence class of $C^\infty(X, \mathbf{T}_L)$ with respect to this equivalence relation. The set of all homotopy classes is denoted by $\pi(X, \mathbf{T}_L)$. Since we have $\phi_1 + \phi_2 \sim \psi_1 + \psi_2$ when $\phi_1 \sim \psi_1$ and $\phi_2 \sim \psi_2$, the set $\pi(X, \mathbf{T}_L)$ is an abelian group.
   If a map ϕ belongs to a homotopy class t, we say that ϕ has a homotopy type t. We also denote the set of maps of homotopy type t by $C^\infty(X, \mathbf{T}_L)_t$.

COROLLARY (5.10).

Let $\mathrm{pr}_*\colon C^\infty(X, \mathfrak{h}_\mathbb{R}) \to C^\infty(X, \mathbf{T}_L)$ be the map given by $\phi \mapsto \mathrm{pr} \circ \phi$, where $\mathrm{pr}\colon \mathfrak{h}_\mathbb{R} \to \mathbf{T}_L$ is the projection map. Then the cokernel $C^\infty(X, \mathbf{T}_L)/\mathrm{pr}_* C^\infty(X, \mathfrak{h}_\mathbb{R})$ is the abelian group of all homotopy classes.

COROLLARY (5.11).

The abelian group of homotopy classes $\pi(X, \mathbf{T}_L)$ is isomorphic to

$$H_1(X; L) \cong L^{\dim H_1(X;\mathbb{Z})}.$$

where H_1 is the first homology group.

The isomorphism $\pi(X, \mathbf{T}_L) \to H_1(X; L)$ is given by

$$\phi \mapsto \frac{1}{2\pi} \left(\int_{\gamma_1} d\phi, \ldots, \int_{\gamma_N} d\phi \right),$$

where $\{\gamma_1, \ldots, \gamma_N\}$ is a basis of the homology group $H_1(X; \mathbb{Z})$ such that each γ_i is a smooth closed path on X.

THEOREM (5.12). [Existence Theorem of Harmonic Maps]

Suppose that X is closed. Then there exists a harmonic map in every homotopy class. Moreover the harmonic maps in the same homotopy classes differ only by additive constants, i.e.,

$$\mathrm{Harm}(X, \mathbf{T}_L) \cong \pi(X, \mathbf{T}_L) \oplus \mathbf{T}_L.$$

Proof : See Eells-Sampson 1964 [ES], where they proved the existence in more general situations. Note that this implies

$$\mathrm{Harm}(X, \mathbf{T}_L)/\mathbf{T}_L \cong H_1(X; L) \cong H^1(X; L),$$

a variant of Hodge theorem.

PROPOSITION (5.13).

When a Riemann surface X is closed we have the following isomorphism

$$C^\infty(X, \mathbf{T}_L) \cong C^\infty{}_*(X, \mathfrak{h}_\mathbb{R}) \oplus \mathbf{T}_L \oplus H_1(X; L).$$

Namely, any map $\phi \in C^\infty(X, \mathbf{T}_L)$ can be written as a sum

$$\phi = \phi_* + \phi_0 + S_\lambda,$$

where $\phi_* \in C^\infty{}_*(X, \mathfrak{h}_\mathbb{R})$, $\phi_0 \in \mathbf{T}_L$, and S_λ is a fixed harmonic function which corresponds to an element λ of the homology group $H_1(X; L)$.

DEFINITION (5.14). [Relative Homotopy Class of Maps]

Suppose that a Riemann surface X is open. Two maps $\phi, \psi \in C^\infty(X, \mathbf{T}_L)$ are said to be homotopic relative to ∂X if there exists a homotopy F connecting ϕ and ψ such that when z is in the boundary ∂X, then $F(t, z)$ does not depend on $t \in I = [0, 1]$. We write $\phi \sim \psi$ (rel ∂X) when ϕ and ψ are homotopic relative to ∂X.

A relative homotopy class of maps from X to $\mathbf{T}_L$ is an equivalence class of $C^\infty(X, \mathbf{T}_L)$ or $C^\infty{}_{X,L}(f_1, \ldots, f_N)$ with respect to this equivalence relation. The set of all relative homotopy classes is denoted by $\pi(X, \partial X; \mathbf{T}_L)$. Since we have $\phi_1 + \phi_2 \sim \psi_1 + \psi_2$ (rel ∂X) when $\phi_1 \sim \psi_1$ (rel ∂X) and $\phi_2 \sim \psi_2$ (rel ∂X), the set $\pi(X, \partial X; \mathbf{T}_L)$ is an abelian group.

If a map ϕ belongs to a relative homotopy class t, we say that ϕ has a relative homotopy type t. We also denote the set of maps in $C^\infty{}_{X,L}(f_1, \ldots, f_N)$ of relative homotopy type t by $C^\infty{}_{X,L}(f_1, \ldots, f_N)_t$. The abelian group of all relative homotopy classes of $C^\infty{}_{X,L}(0, \ldots, 0)$ is denoted by $\pi(X, \partial X; \mathbf{T}_L, 0)$.

COROLLARY (5.15).
$$\pi(X, \partial X; \mathbf{T}_L) \cong C^\infty(X, \mathbf{T}_L)/C^\infty{}_{X,L}(0, \ldots 0)_0.$$

COROLLARY (5.16).

The abelian group $\pi(X, \partial X; \mathbf{T}_L, 0)$ of all relative homotopy classes of $C^\infty{}_{X,L}(0, \ldots, 0)$ is isomorphic to
$$H_1(X, \partial X; L) \cong L^{\dim H_1(X, \partial X; \mathbb{Z})}.$$
where H_1 is the first relative homology group

The isomorphism $\pi(X, \partial X; \mathbf{T}_L, 0) \to H_1(X, \partial X; L)$ is given by
$$\phi \mapsto \frac{1}{2\pi} \left(\int_{\gamma_1} d\phi, \ldots, \int_{\gamma_N} d\phi \right) + c,$$
where $\{\gamma_1, \ldots, \gamma_N\}$ is a basis of $H_1(X, \partial X; \mathbb{Z})$ such that each γ_i is either a smooth closed path on X or a smooth open path on X starting and ending at the boundary of X, and $c \in \mathfrak{h}_{\mathbb{R}}{}^N$ is an appropriate constant depending on the choice of the basis.

PROPOSITION (5.17). [Solvability of Dirichlet Problem]

Suppose that a Riemann surface X is open. There exists a unique harmonic map for every relative homotopy class. Namely,
$$\mathrm{Harm}(X, \mathbf{T}_L) \cong \pi(X, \partial X; \mathbf{T}_L)$$
More specificly, for maps on the boundary $f_1, \ldots, f_N \in C^\infty(\mathbf{S}^1, \mathbf{T}_L)$, put $f_i = f_{i*} + f_{i0} + \lambda_i \theta$. Define $\lambda \in L$ by $\lambda = \sum_{i=1}^N \pm \lambda_i$ where $\pm$ is defined such that $+$ for out-circles and $-$ for in-circles.

If $\lambda \neq 0$, then $C^\infty{}_{X,L}(f_1, \ldots, f_N)$ is empty.

If $\lambda = 0$, then there exists a unique harmonic map ϕ_0 on X such that

$$\phi_0 = (f_1, \ldots, f_N), \quad \text{on } \partial X,$$

for every relative homotopy class t of $C^\infty{}_{X,L}(f_1, \ldots, f_N)$. We denote the above unique harmonic map ϕ_0 by $\phi_X(f_1, \ldots, f_N)_t$.

Note that we have

$$\phi_X(c, \ldots, c)_0 = c \quad \text{for } c : \text{a constant},$$

$$\phi_X(f_1, \ldots, f_N)_t + \phi_X(g_1, \ldots, g_N)_s = \phi_X(f_1 + g_1, \ldots, f_N + g_N)_{t+s}.$$

Proof : See Hamilton 1975 [Ham], where he proved the existence in more general situations. The uniqueness can be easily proved.

PROPOSITION **(5.18)**. [Green Orthogonal Decomposition]
When the Riemann surface X is open, by the Green's theorem and the solvability of Dirichlet problem, we have an orthogonal decomposition with respect to the Dirichlet inner product $\ll , \gg$

$$C^\infty(X, \mathbf{T}_L) \cong C^\infty{}_{X,L}(0, \ldots, 0)_0 \oplus \mathrm{Harm}(X, \mathbf{T}_L).$$

Proof : First, note that $C^\infty{}_{X,L}(0, \ldots, 0)_0$ can be identified with $C^\infty{}_X(0, \ldots, 0)$. Namely any map $\phi \in C^\infty{}_{X,L}(0, \ldots, 0)$ can be lifted to $\phi \in C^\infty{}_X(0, \ldots, 0)$ uniquely.

Assume that $\phi \in C^\infty{}_{X,L}(0, \ldots, 0)$ and $\Delta_g \varphi = 0$. Then the Green's theorem states that we have

$$\ll \phi, \varphi \gg = \int_{\partial X} \phi * d\varphi - \iint_X <\phi, \Delta_g \varphi>_\Gamma \omega_g = 0,$$

where $*d\varphi = -\varphi_y dx + \varphi_x dy$, locally. Here we used the lifting of ϕ.

Also for $\phi \in C^\infty{}_{X,L}(f_1, \ldots, f_N)_t$, the decomposition into non-harmonic and harmonic part is given by

$$\phi \mapsto (\phi - \phi_0, \phi_0),$$

where $\phi_0 = \phi_X(f_1, \ldots, f_N)_t$.

COROLLARY **(5.19)**.
Let X be an open Riemann surface. Then for each relative homotopy type t, we have

$$C^\infty{}_{X,L}(f_1, \ldots, f_N)_t \cong C^\infty{}_{X,L}(0, \ldots, 0)_0 : \phi + \phi_0 \leftrightarrow \phi,$$

where $\phi_0 = \phi_X(f_1, \ldots, f_N)_t$.

§5-B. Examples.

We would like to examine several elementary examples for which explicit computations are possible. We will use these computations to construct the action of vertex operators explicitly.

Example **(5.20).** [Disc]

Let $D_q = \{z \in \mathbb{C} \mid |z| \leq |q|\} \subset \mathbb{C}$ be the disc of radius $= |q|$, with the boundary parametrized by

$$\mathbf{S}^1 \to \partial D_q : \theta \mapsto qe^{i\theta}, \qquad \text{(out-circle)}$$

and with the metric $dz\,d\bar{z}$ induced by the complex plane $\mathbb{C}$.

We have $H_1(D_q; L) = O$, $H_1(D_q, \partial D_q; L) = O$. Also we have $Area(D_q) = \pi q\bar{q}$.

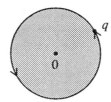

(1) For a function $f(\theta) = \sum_{n \in \mathbb{Z}} f_n e^{in\theta} \in C^\infty(\mathbf{S}^1, \mathfrak{h}_\mathbb{R})$ on the boundary, the unique harmonic function $\widehat{f} = \phi_{D_q}(f) \in \mathrm{Harm}(D_q, \mathfrak{h}_\mathbb{R}) \cap C^\infty{}_{D_q}(f)$ is given by

$$\widehat{f}(z) = f_0 + \sum_{n=1}^{\infty}\left(f_n\frac{z^n}{q^n} + \bar{f}_n\frac{\bar{z}^n}{\bar{q}^n}\right).$$

We have

$$I_{D_q}(\widehat{f}) = \sum_{n=1}^{\infty} n <f_n, \bar{f}_n>_\Gamma = \frac{1}{2}(f_*, \bar{f}_*).$$

Note that since the action integral I is conformally invariant, it does not depend on the parameter q.

(2) Let $f = f_* + f_0 + \lambda\theta \in C^\infty(\mathbf{S}^1, \mathbf{T}_L)$ be a map on the boundary.

If $\lambda \neq 0$, then the set $C^\infty{}_{D_q, L}(f)$ is empty.

On the other hand, if $\lambda = 0$, then the unique harmonic map $\widehat{f} = \phi_{D_q}(f) \in \mathrm{Harm}_{D_q, L}(f)$ is given by the same formula

$$\widehat{f}(z) = f_0 + \sum_{n=1}^{\infty}\left(f_n\frac{z^n}{q^n} + \bar{f}_n\frac{\bar{z}^n}{\bar{q}^n}\right).$$

We have

$$I_{D_q}(\widehat{f}) = \sum_{n=1}^{\infty} n <f_n, \bar{f}_n>_\Gamma = \frac{1}{2}(f_*, \bar{f}_*).$$

Note that D_q is homotopically trivial, and there exists only one relative homotopy class of $C^\infty{}_{D_q, L}(f)$.

Example **(5.21)**. [Annulus and Cylinder]

(I) Annulus.

Let q' and q'' be two complex numbers such that $|q''| < |q'|$, and let

$$A_{q',q''} = \{z \in \mathbb{C} \mid |q''| \le |z| \le |q'|\} \subset \mathbb{C}$$

be the annulus of the outer radius $= |q'|$ and the inner radius $= |q''|$ with the parametrization of the boundary given by

$$\mathbf{S}^1 \to \partial A_{q',q''} : \theta \mapsto q'e^{i\theta} \qquad \text{(out-circle)}$$

$$\mathbf{S}^1 \to \partial A_{q',q''} : \theta \mapsto q''e^{i\theta} \qquad \text{(in-circle)}$$

and with the metric $dzd\bar{z}$ induced by $\mathbb{C}$. Then we have $Area(A_{q',q''}) = \pi(q'\overline{q'} - q''\overline{q''})$.

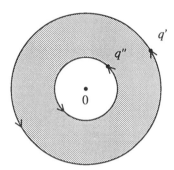

(II) Cylinder.

Let $\tau' = \tau'_1 + i\tau'_2$ and $\tau'' = \tau''_1 + i\tau''_2$ be two complex numbers such that $\tau''_2 > \tau'_2$, and let

$$C_{\tau',\tau''} = \{u \in \mathbb{C} \mid 2\pi\tau'_2 \le \mathrm{Im}\,u \le 2\pi\tau''_2\}/2\pi\mathbb{Z}$$

be the cylinder with the parametrization of the boundary given by

$$\mathbf{S}^1 \to \partial C_{\tau',\tau''} : \theta \mapsto \theta + 2\pi\tau' \qquad \text{(out-circle)}$$

$$\mathbf{S}^1 \to \partial C_{\tau',\tau''} : \theta \mapsto \theta + 2\pi\tau'' \qquad \text{(in-circle)}$$

and with the metric $dud\bar{u} = dx^2 + dy^2$ induced by the covering $\mathbb{C}$. Here $u = x + iy$. Then $Area(C_{\tau',\tau''}) = 4\pi^2(\tau''_2 - \tau'_2)$.

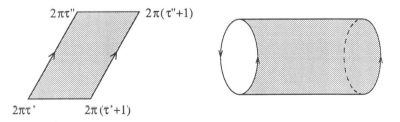

Note that the cylinder $C_{\tau',\tau''}$ is conformally equivalent to the annulus $A_{q',q''}$ when $q' = e^{2\pi i\tau'}$ and $q'' = e^{2\pi i\tau''}$, with the mapping

$$C_{\tau',\tau''} \to A_{q',q''} : u \mapsto z = e^{iu},$$

but the metrics are not equal.

We state several results which do not depend on the metric. In the following, X means either $A_{q',q''}$ or $C_{\tau',\tau''}$. z is the standard coordinate of $A_{q',q''}$ and $u = x + iy$ is the standard coordinate of $C_{\tau',\tau''}$. We also put $q = q''/q'$ and $\tau = \tau'' - \tau'$.

We have $H_1(X;L) = L$, $H_1(X, \partial X; L) = L$.

(1) For two functions $f(\theta) = \sum_{n\in\mathbb{Z}} f_n e^{in\theta}$ and $g(\theta) = \sum_{n\in\mathbb{Z}} g_n e^{in\theta} \in C^\infty(\mathbf{S}^1, \mathfrak{h}_{\mathbb{R}})$ on the boundary, the unique harmonic function $\phi_0 = \phi_X(f,g) \in \mathrm{Harm}(X, \mathfrak{h}_{\mathbb{R}}) \cap C^\infty{}_X(f,g)$ is given by

$$\phi_0(z) = f_0 + (g_0 - f_0)\frac{y'}{2\pi\tau_2}$$

$$+\sum_{n=1}^\infty \frac{1}{1-q^n\overline{q}^n}\left\{ f_n(z'^n - q^n\overline{q}^n\overline{z'}^{-n}) + \overline{f}_n(\overline{z'}^n - q^n\overline{q}^n z'^{-n}) \right.$$

$$\left. + g_n\overline{q}^n(\overline{z'}^{-n} - z'^n) + \overline{g}_n q^n(z'^{-n} - \overline{z'}^n)\right\}.$$

where $z' = z/q', y' = y - 2\pi\tau_2'$. Namely, ϕ_0 satisfies

$$\phi_0(q'e^{i\theta}) = f(\theta) \quad \text{and} \quad \phi_0(q''e^{i\theta}) = g(\theta).$$

We have

$$I_X(\phi_0) = \frac{1}{4\pi\tau_2} < f_0 - g_0, f_0 - g_0 >_\Gamma$$

$$+\sum_{n=1}^\infty \frac{n}{1-q^n\overline{q}^n}\left((1+q^n\overline{q}^n)(<f_n, \overline{f}_n>_\Gamma + <g_n, \overline{g}_n>_\Gamma) \right.$$

$$\left. -2q^n <f_n, \overline{g}_n>_\Gamma - 2\overline{q}^n <\overline{f}_n, g_n>_\Gamma \right)$$

$$= \frac{1}{4\pi\tau_2} < f_0 - g_0, f_0 - g_0 >_\Gamma$$

$$+\frac{1}{2}(f_*, \overline{f}_*) + \frac{1}{2}(g_*, \overline{g}_*) + \left(\frac{Tf_*}{\sqrt{1-T\overline{T}}}, \frac{\overline{Tf_*}}{\sqrt{1-T\overline{T}}} \right) - \left(\frac{Tf_*}{\sqrt{1-T\overline{T}}}, \frac{\overline{g}_*}{\sqrt{1-T\overline{T}}} \right)$$

$$-\left(\frac{g_*}{\sqrt{1-T\overline{T}}}, \frac{\overline{Tf_*}}{\sqrt{1-T\overline{T}}} \right) + \left(\frac{Tg_*}{\sqrt{1-T\overline{T}}}, \frac{\overline{Tg_*}}{\sqrt{1-T\overline{T}}} \right),$$

where

$$T = T_q : C^\infty{}_*(\mathbf{S}^1, \mathfrak{h}_{\mathbb{R}}) \to C^\infty{}_*(\mathbf{S}^1, \mathfrak{h}_{\mathbb{R}})$$

is the operator given by

$$T_q f_* = \sum_{n=1}^\infty f_n q^n e^{in\theta} + \sum_{n=1}^\infty \overline{f}_n \overline{q}^n e^{-in\theta}.$$

Note that since the action integral I is conformally invariant, it depends only on q or τ.

In particular, for a function $f(\theta) = \sum_{n\in\mathbb{Z}} f_n e^{in\theta} \in C^\infty(\mathbf{S}^1, \mathfrak{h}_{\mathbb{R}})$, the unique harmonic function $\widetilde{f} = \phi_X(f,f) \in \mathrm{Harm}(X, \mathfrak{h}_{\mathbb{R}}) \cap C^\infty{}_X(f,f)$ is given by

$$\widetilde{f}(z) = f_0 + \sum_{n=1}^\infty \frac{1}{1-q^n\overline{q}^n}\Big\{ f_n\{(1-\overline{q}^n)z'^n + \overline{q}^n(1-q^n)\overline{z}'^{-n}\}$$
$$+\overline{f}_n\{(1-q^n)\overline{z}'^n + q^n(1-\overline{q}^n)z'^{-n}\}\Big\}.$$

where $z' = z/q'$. We have

$$I_X(\widetilde{f}) = \sum_{n=1}^\infty 2n\frac{(1-q^n)(1-\overline{q}^n)}{1-q^n\overline{q}^n}<f_n,\overline{f}_n>_\Gamma= \left(\frac{(1-T)f_*}{\sqrt{1-T\overline{T}}}, \frac{\overline{(1-T)f_*}}{\sqrt{1-T\overline{T}}}\right),$$

where $T = T_q$.

(2) Let $f = f_* + f_0 + \lambda\theta$, and $g = g_* + g_0 + \mu\theta$ be two maps in $C^\infty(\mathbf{S}^1, \mathbf{T}_L)$

When $\lambda \neq \mu$, there is no element in $C^\infty{}_{X,L}(f,g)$.

On the other hand, when $\lambda = \mu$, an element of $C^\infty{}_{X,L}(f,g)$ can be written as

$$\phi_* + f_0 + S_{\lambda,\beta+(g_0-f_0)/2\pi},$$

where $\phi_* \in C^\infty{}_X(f_*,g_*)$ and

$$S_{\lambda,\beta+(g_0-f_0)/2\pi}(z) = \lambda\Big((x-2\pi\tau_1') - \frac{\tau_1}{\tau_2}(y-2\pi\tau_2')\Big) + \Big(\beta+\frac{g_0-f_0}{2\pi}\Big)\frac{y-2\pi\tau_2'}{\tau_2},$$

$\beta \in L$. Note that β corresponds to a relative homotopy class. We have

$$C^\infty{}_{X,L}(f,g) = \begin{cases} C^\infty{}_X(f_*,g_*) \oplus L, & \text{if } \lambda = \mu \\ \emptyset, & \text{if } \lambda \neq \mu. \end{cases}$$

The harmonic maps in $C^\infty{}_{X,L}(f,g)$ are of form

$$\phi_{0*}(z) + f_0 + S_{\lambda,\beta+(g_0-f_0)/2\pi},$$

where $\phi_{0*} = \phi_X(f_*,g_*)$ and $\beta \in L$.

In particular, for a map $f = f_* + f_0 + \lambda\theta \in C^\infty(\mathbf{S}^1,\mathbf{T}_L)$, the harmonic maps in $C^\infty{}_{X,L}(f,f)$ are given by

$$\widetilde{f}_*(z) + f_0 + S_{\lambda,\beta},$$

where $\beta \in L$.

Example **(5.22).** [N-holed disk]

Let $P = P_{z_1,\rho_1,\ldots,z_N,\rho_N;q}$ be the N-holed disk

$$D_q - \left(D_{z_1,\rho_1} \cup \cdots \cup D_{z_N,\rho_N}\right)^\circ,$$

where $D_{z,\rho}$ is the disk of radius$= |\rho|$ centered at z, and ( )$^\circ$ means the interior.

We use the parametrization of the boundary

$$\mathbf{S}^1 \to \partial P : \theta \mapsto z_1 + \rho_1 e^{i\theta} \qquad \text{(in-circle)}$$
$$\vdots$$
$$\mathbf{S}^1 \to \partial P : \theta \mapsto z_N + \rho_N e^{i\theta} \qquad \text{(in-circle)}$$
$$\mathbf{S}^1 \to \partial P : \theta \mapsto q e^{i\theta} \qquad \text{(out-circle)}$$

and the metric $dz d\bar{z}$ induced by $\mathbb{C}$. We have $H_1(P;L) = L^N$, $H_1(P,\partial P;L) = L^N$.

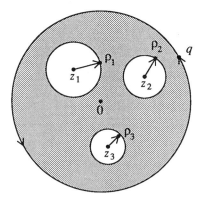

Even in this rather simple example, it is impossible to write down the concrete formula which express the correspondence between harmonic functions and the functions on the boundary as in the case of cylinder.

Example **(5.23).** [Elliptic Curve]

Let $\tau = \tau_1 + i\tau_2$ be a complex number such that $\tau_2 > 0$, and let

$$E_\tau = \mathbb{C}/(2\pi\mathbb{Z} + 2\pi\tau\mathbb{Z})$$

be the elliptic curve defined by τ. We use the standard metric $du d\bar{u} = dx^2 + dy^2$, where $u = x + iy$. We also use a coordinate $z = e^{iu}$. We have $H_1(E_\tau;L) = L^2$. Also $Area(E_\tau) = 4\pi^2\tau_2$.

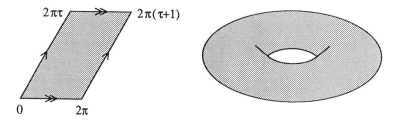

(1) Any function ϕ in $C^\infty(E_\tau, \mathfrak{h}_{\mathbb{R}})$ can be expanded as

$$\phi(z) = \sum_{n,m \in \mathbb{Z}} \phi_{n,m} e^{i[n(x - \frac{\tau_1}{\tau_2}y) - m\frac{y}{\tau_2}]}, \quad \phi_{n,m} \in \mathfrak{h}, \quad \phi_{-n,-m} = \overline{\phi}_{n,m}.$$

Then $C^\infty{}_*(E_\tau, \mathfrak{h}_{\mathbb{R}})$ is the space of C^∞-functions on E_τ without constant term $\phi_{0,0}$. Note that $C^\infty{}_*(E_\tau, \mathfrak{h}_{\mathbb{R}})$ has a natural $\mathbb{C}$-vector space structure on it.

(2) Any map $\phi \in C^\infty(E_\tau, \mathbf{T}_L)$ can be written as

$$\phi = \phi_* + \phi_{0,0} + S_{\alpha,\beta},$$

where $\phi_* \in C^\infty{}_*(E_\tau, \mathfrak{h}_{\mathbb{R}})$, $\phi_{0,0} \in \mathbf{T}_L$ and

$$S_{\alpha,\beta}(z) = \alpha \left(x - \frac{\tau_1}{\tau_2} y \right) + \beta \frac{y}{\tau_2}, \quad \alpha, \beta \in L.$$

Each $S_{\alpha,\beta}$ corresponds to a homotopy class of $C^\infty(E_\tau, \mathbf{T}_L)$. Therefore we have an isomorphism of abelian groups

$$C^\infty(E_\tau, \mathbf{T}_L) \cong C^\infty{}_*(E_\tau, \mathfrak{h}_{\mathbb{R}}) \oplus \mathbf{T}_L \oplus (L \times L).$$

(See Proposition (5.13).)

§6. Geometric Realization of Vertex Operators.

For any Riemann surface X with boundary consisting of N circles, and for any functions $f_1, \ldots, f_N : \mathbf{S}^1 \to \mathfrak{h}_{\mathbb{R}}$ on the components of the boundary, a formal expression of a string path integral

$$K_X(f_1, \ldots, f_N) = \int_{C^\infty{}_X(f_1, \ldots, f_N)} e^{-I_X(\phi)}[d\phi],$$

should be defined so that we have

$$K_X(f_1, \ldots, f_N) = K_X(0, \ldots, 0)\, e^{-I_X(\phi_0)},$$

where $\phi_0 = \phi_X(f_1, \ldots, f_N)$. This is because any element of the space $C^\infty{}_X(f_1, \ldots, f_N)$ can be written as a sum $\phi + \phi_0$, where $\phi \in C^\infty{}_X(0, \ldots, 0)$, and we have

$$I_X(\phi + \phi_0) = I_X(\phi) + I_X(\phi_0).$$

(See Corollary (5.8).) Similarly, for any maps $f_1, \ldots, f_N : \mathbf{S}^1 \to \mathbf{T}_L$ on the components of the boundary, a formal expression of a string path integral

$$K_{X,L}(f_1, \ldots, f_N) = \int_{C^\infty{}_{X,L}(f_1, \ldots, f_N)} e^{-I_X(\phi)}[d\phi],$$

should be defined so that we have

$$K_{X,L}(f_1, \ldots, f_N) = K_X(0, \ldots, 0) \sum_{\phi_0} e^{-I_X(\phi_0)},$$

where the sum is over all harmonic maps $\phi_0 \in \mathrm{Harm}_{X,L}(f_1, \ldots, f_N)$. Again this is because any element of the space $C^\infty{}_{X,L}(f_1, \ldots, f_N)$ can be written uniquely as a sum $\phi + \phi_0$, where $\phi \in C^\infty{}_X(0, \ldots, 0)$, and $\phi_0 \in \mathrm{Harm}_{X,L}(f_1, \ldots, f_N)$, and we have

$$I_X(\phi + \phi_0) = I_X(\phi) + I_X(\phi_0).$$

(See Corollary (5.19).)

Therefore, to define $K_X(f_1, \ldots, f_N)$ and $K_{X,L}(f_1, \ldots, f_N)$, all we have to do is fix a positive constant

$$K_X = K_X(0, \ldots, 0).$$

In this section we restrict our attention to N-holed disks P. Note that they are closed under sewing. We define all the constants K_P so that the Markov property is satisfied.

Namely if P_3 is obtained by sewing P_1 and P_2, then we have

(1)
$$\int_{C^\infty(\mathbf{S}^1,\mathfrak{h}_{\mathbb{R}})^\wedge} K_{P_1}(f,f_1,\dots,f_N,g)\, K_{P_2}(g,h_1,\dots,h_M)[dg]$$
$$= K_{P_3}(f,f_1,\dots,f_N,h_1,\dots,h_M),$$

(2)
$$\int_{C^\infty(\mathbf{S}^1,\mathbf{T}_L)^\wedge} K_{P_1,L}(f,f_1,\dots,f_N,g)\, K_{P_2,L}(g,h_1,\dots,h_M)[dg]$$
$$= K_{P_3,L}(f,f_1,\dots,f_N,h_1,\dots,h_M).$$

(The use of a symbol $[dg]$ is not standard. The precise meaning of these formulas will be explained below.)

With this definition of string path integrals, we show that the action of the neutral vertex operators can be realized by string path integrals on 2-holed disks. By sewing two of such 2-holed disks, we can prove the associative law of the vertex operators geometrically.

§6-A. Gaussian Integrals over Harmonic Functions.

PROPOSITION (6.1).

Let q', q'' be two complex numbers such that $|q''| < |q'|$. We put $q = q''/q'$. Let $A = A_{q',q''}$ be an annulus defined in Example (5.21).

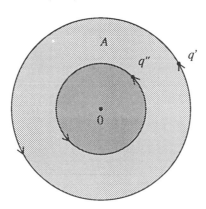

Note that the disk $D_{q'}$ is obtained by sewing A and $D_{q''}$. Let $\mathrm{Harm}(A, D_{q''})$ be the $\mathbb{R}$-vector space of functions ϕ on $D_{q'}$ with values in $\mathfrak{h}_{\mathbb{R}}$ such that ϕ is harmonic on A and on $D_{q''}$ and also $\phi|_{\partial D_{q'}} = 0$. Since any function $\phi \in \mathrm{Harm}(A, D_{q''})$ can be uniquely determined by its values on the circle $\partial D_{q''} = D_{q''} \cap A$, we have an isomorphism

$$C^\infty(\mathbf{S}^1,\mathfrak{h}_{\mathbb{R}}) \cong \mathrm{Harm}(A, D_{q''}) : g \mapsto \widetilde{g},$$

where

$$\widetilde{\widetilde{g}} = \begin{cases} \phi_A(0,g), & \text{on } A, \\ \widehat{g}, & \text{on } D_{q''}. \end{cases}$$

We can compute $I(\widetilde{g})$ explicitly.

$$I(\widetilde{g}) = I_D(\widehat{g}) + I_A(\phi_A(0,g)) = \frac{1}{4\pi\tau_2} <g_0, g_0 >_\Gamma + \left(\frac{1}{\sqrt{1-T\overline{T}}} g_*, \frac{1}{\sqrt{1-T\overline{T}}} \overline{g}_* \right),$$

where $T = T_q$ is the operator defined in Example (5.21), and $q = e^{2\pi i \tau}$, $\tau = \tau_1 + i\tau_2$. Note that the operator

$$B = \frac{1}{\sqrt{1-T\overline{T}}}$$

is of form $Id + \text{traceclass}$.

PROPOSITION (6.2).
We have

$$\int_{C^\infty(\mathbf{S}^1, \mathfrak{h}_\mathbb{R})^\wedge} [e^{-I(\widetilde{g})} dg] = \sqrt{2\tau_2}^\ell \cdot \det(1 - T\overline{T}) = \left(\sqrt{2\tau_2} f(q\overline{q}) \right)^\ell.$$

where $f(q) = \prod_{n=1}^\infty (1 - q^n)$.

PROPOSITION (6.3).
Let q'', q''', q'''' be three complex numbers such that $|q''''| < |q'''| < |q''|$. We put $q = q'''/q''$ and $q' = q''''/q'''$. Let $A_1 = A_{q'',q'''}$ and $A_2 = A_{q''',q''''}$ be two annuluses defined in Example (5.21).

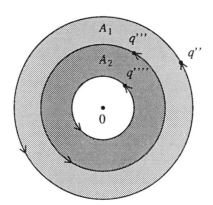

Let $A_3 = A_{q'',q''''}$ be the annulus obtained by sewing the inner circle boundary of A_1 and the outer boundary of A_2. Let $\text{Harm}(A_1, A_2)$ be the $\mathbb{R}$-vector space of functions ϕ on A_3 with values in $\mathfrak{h}_\mathbb{R}$ such that ϕ is harmonic on A_1 and on A_2 and also $\phi|_{\partial A_3} = 0$. Since any function $\phi \in \text{Harm}(A_1, A_2)$ can be uniquely determined by its values on the circle $\partial D_{q'''} = A_1 \cap A_2$, we have an isomorphism

$$C^\infty(\mathbf{S}^1, \mathfrak{h}_\mathbb{R}) \cong \text{Harm}(A_1, A_2) : g \mapsto \widetilde{\widetilde{g}},$$

where

$$\widetilde{\widetilde{g}} = \begin{cases} \phi_{A_1}(0,g), & \text{on } A_1, \\ \phi_{A_2}(g,0), & \text{on } A_2. \end{cases}$$

We can compute $I_{A_3}(\widetilde{\widetilde{g}})$ explicitly.

$$I_{A_3}(\widetilde{\widetilde{g}}) = I_{A_1}(\phi_{A_1}(0,g)) + I_{A_2}(\phi_{A_2}(g,0))$$

$$= \frac{1}{4\pi}\frac{\tau_2 + \tau_2'}{\tau_2\tau_2'} <g_0, g_0>_\Gamma + \left(\sqrt{\frac{1 - TT'\overline{T}\overline{T'}}{(1 - T\overline{T})(1 - T'\overline{T'})}}\,g_*, \overline{\sqrt{\frac{1 - TT'\overline{T}\overline{T'}}{(1 - T\overline{T})(1 - T'\overline{T'})}}\,g_*} \right),$$

where $T = T_q$ and $T = T_{q'}$ are the operator defined in Example (5.21). Note that the operator

$$B = \sqrt{\frac{1 - TT'\overline{T}\overline{T'}}{(1 - T\overline{T})(1 - T'\overline{T'})}}$$

is of form $Id + \text{traceclass}$.

PROPOSITION (6.4).
We have

$$\int_{C^\infty(\mathbf{S}^1, \mathfrak{h}_\mathbb{R})^\wedge} [e^{-I(\widetilde{\widetilde{g}})}dg]$$

$$= \sqrt{\frac{2\tau_2\tau_2'}{\tau_2 + \tau_2'}}^\ell \left(\det\frac{1 - TT'\overline{T}\overline{T'}}{(1 - T\overline{T})(1 - T'\overline{T'})} \right)^{-1} = \left(\frac{\sqrt{2\tau_2}\sqrt{2\tau_2'}}{\sqrt{2(\tau_2 + \tau_2')}} \frac{f(q\overline{q})f(q'\overline{q'})}{f(qq'\overline{qq'})} \right)^\ell.$$

PROPOSITION (6.5).
Let P_1 be an $(N+1)$-holed disk and P_2 be an M-holed disk. Let P_3 be the $(N+M)$-holed disk obtained by sewing an inner circle boundary of P_1 and the outer circle boundary of P_2.

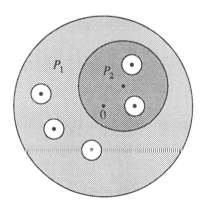

For any function $g \in C^\infty(\mathbf{S}^1, \mathfrak{h}_\mathbb{R})$ on $P_1 \cap P_2$, we define a function $\widetilde{\widetilde{g}}$ on P_3 by

$$\widetilde{\widetilde{g}} = \begin{cases} \phi_{P_1}(0, \ldots, 0, g), & \text{on } P_1, \\ \phi_{P_2}(g, 0, \ldots, 0), & \text{on } P_2. \end{cases}$$

We have the following estimate about $I_{P_3}(\widetilde{\widetilde{g}})$. Take two annuluses A_1 and A_2 such that

$$A_1 \subset P_1 \subset A_2,$$

with the same inner circle boundary.

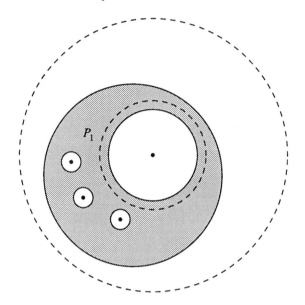

Then because the harmonic function minimizes the action I, we have

$$I_{A_1}(\phi_{A_1}(0, g)) \geq I_{P_1}(\phi_{P_1}(0, \ldots, 0, g)) \geq I_{A_2}(\phi_{A_2}(0, g)).$$

Take an annulus A_3 and the disk D such that

$$A_3 \subset P_2 \subset D,$$

with the same outer circle boundary.

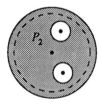

Then again because the harmonic function minimizes the action I, we have

$$I_{A_3}(\phi_{A_3}(g, 0)) \geq I_{P_1}(\phi_{P_1}(g, 0, \ldots, 0)) \geq I_D(\widehat{g}).$$

Therefore we have

$$e^{-I_{A_1 \cup A_3}(\widetilde{\widetilde{g}})} \leq e^{-I_{P_3}(\widetilde{\widetilde{g}})} \leq e^{-I_{A_2 \cup D}(\widetilde{\widetilde{g}})}.$$

This estimate and Propositions (6.1) and (6.3) implies that the Gaussian integral

$$\int_{C^\infty(\mathbf{S}^1, \mathfrak{h}_{\mathbf{R}})^\wedge} [e^{-I_{P_3}(\widetilde{\widetilde{g}})} dg]$$

has a finite positive value.

PROPOSITION (6.6). [Sewing of Two Disks with Holes]
Under the same assumption of Proposition (6.5), for functions $f, f_1, \ldots, f_N \in C^\infty(\mathbf{S}^1, \mathfrak{h}_{\mathbf{R}})$ on the P_1-side of the boundary of P_3, and functions $h_1, \ldots, h_M \in C^\infty(\mathbf{S}^1, \mathfrak{h}_{\mathbf{R}})$ on the P_2-side of the boundary of P_3, we have

$$\int_{C^\infty(\mathbf{S}^1, \mathfrak{h}_{\mathbf{R}})^\wedge} e^{-I_{P_1}(\phi_0)} e^{-I_{P_2}(\psi_0)} [dg] = \int_{C^\infty(\mathbf{S}^1, \mathfrak{h}_{\mathbf{R}})^\wedge} [e^{-I_{P_3}(\widetilde{\widetilde{g}})} dg] \cdot e^{-I_{P_3}(\varphi_0)},$$

where
  g is a function on $S = P_1 \cap P_2$,
  $\phi_0 = \phi_{P_1}(f, f_1, \ldots, f_N, g)$,
  $\psi_0 = \phi_{P_2}(g, h_1, \ldots, h_M)$,
  $\varphi_0 = \phi_{P_3}(f, f_1, \ldots, f_N, h_1, \ldots, h_M)$,
  and $\widetilde{\widetilde{g}} = \begin{cases} \phi_{P_1}(0, 0, \ldots, 0, g), & \text{on } P_1 \\ \phi_{P_2}(g, 0, \ldots, 0), & \text{on } P_2. \end{cases}$

Proof : Given functions $f, f_1, \ldots, f_N, h_1, \ldots, h_M \in C^\infty(\mathbf{S}^1, \mathfrak{h}_{\mathbf{R}})$, let us define a function $k \in C^\infty(\mathbf{S}^1, \mathfrak{h}_{\mathbf{R}})$ by

$$k = \varphi_0\big|_S.$$

Then by the uniqueness of harmonic function, we have

$$\varphi_0 = \begin{cases} \phi_{P_1}(f, f_1, \ldots, f_N, k), & \text{on } P_1 \\ \phi_{P_2}(k, h_1, \ldots, h_M), & \text{on } P_2. \end{cases}$$

Now given a function $g \in C^\infty(\mathbf{S}^1, \mathfrak{h}_{\mathbf{R}})$ on S, define $\Phi_0 \in C^\infty(P_3, \mathfrak{h}_{\mathbf{R}})$ by

$$\Phi_0 = \begin{cases} \phi_0, & \text{on } P_1 \\ \psi_0, & \text{on } P_2. \end{cases}$$

Then we have

$$\Phi_0 = \varphi_0 + \widetilde{\widetilde{g - k}}.$$

Moreover we have

$$I_{P_3}(\Phi_0) = I_{P_3}(\varphi_0) + I_{P_3}(\widetilde{\widetilde{g - k}}).$$

since $\widetilde{\widetilde{g - k}} = 0$ on the boundary of P_3.

Therefore we get,

$$\int_{C^\infty(\mathbf{S}^1, \mathfrak{h}_\mathbf{R})^\wedge} e^{-I_{P_1}(\phi_0)} e^{-I_{P_2}(\psi_0)} [dg] = \int_{C^\infty(\mathbf{S}^1, \mathfrak{h}_\mathbf{R})^\wedge} [e^{-I_{P_3}(\Phi_0)} dg]$$

$$= \int_{C^\infty(\mathbf{S}^1, \mathfrak{h}_\mathbf{R})^\wedge} [e^{-I_{P_3}(\widetilde{\widetilde{g-k}})} dg] \cdot e^{-I_{P_3}(\varphi_0)} = \int_{C^\infty(\mathbf{S}^1, \mathfrak{h}_\mathbf{R})^\wedge} [e^{-I_{P_3}(\widetilde{\widetilde{g}})} dg] \cdot e^{-I_{P_3}(\varphi_0)}.$$

DEFINITION (6.7). [Sewing of Two Relative Homotopy Classes]
Under the same assumption of Proposition (6.5), let $f, f_1, \ldots, f_N \in C^\infty(\mathbf{S}^1, \mathbf{T}_L)$ be maps on the P_1-side of the boundary of P_3, and let $h_1, \ldots, h_M \in C^\infty(\mathbf{S}^1, \mathbf{T}_L)$ be maps on the P_2-side of the boundary of P_3. Also let $g \in C^\infty(\mathbf{S}^1, \mathbf{T}_L)$ be a map on $S = P_1 \cap P_2$.

Let t be a relative homotopy class of $C^\infty{}_{P_1, L}(f, f_1, \ldots, f_N, g)$ and let s be a relative homotopy class of $C^\infty{}_{P_2, L}(g, h_1, \ldots, h_M)$. They determine a unique relative homotopy class u of $C^\infty{}_{P_3, L}(f, f_1, \ldots, f_N, h_1, \ldots, h_M)$. We denote this relative homotopy class u by $t * s$. Namely, if $\phi \in C^\infty{}_{P_1, L}(f, f_1, \ldots, f_N, g)_t$ and $\psi \in C^\infty{}_{P_2, L}(g, h_1, \ldots, h_M)_s$ then the function

$$\varphi \in C^\infty{}_{P_3, L}(f, f_1, \ldots, f_N, h_1, \ldots, h_M)$$

such that

$$\varphi = \begin{cases} \phi, & \text{on } P_1 \\ \psi, & \text{on } P_2, \end{cases}$$

has a relative homotopy type $t * s$.

PROPOSITION (6.8).
We have

$$\int_{C^\infty(\mathbf{S}^1, \mathbf{T}_L)^\wedge} \sum_t e^{-I_{P_1}(\phi_{0t})} \sum_s e^{-I_{P_2}(\psi_{0s})} [dg] = \int_{C^\infty(\mathbf{S}^1, \mathfrak{h}_\mathbf{R})^\wedge} [e^{-I_{P_3}(\widetilde{\widetilde{g}})} dg] \cdot \sum_u e^{-I_{P_3}(\varphi_{0u})},$$

where

t is a relative homotopy class of $C^\infty{}_{P_1, L}(f, f_1, \ldots, f_N, g)$
$$\text{and } \phi_{0t} = \phi_{P_1}(f, f_1, \ldots, f_N, g)_t,$$
s is a relative homotopy class of $C^\infty{}_{P_2, L}(g, h_1, \ldots, h_M)$ and $\psi_{0s} = \phi_{P_2}(g, h_1, \ldots, h_M)_s,$
u is a relative homotopy class of $C^\infty{}_{P_3, L}(f, f_1, \ldots, f_N, h_1, \ldots, h_M)$
$$\text{and } \varphi_{0u} = \phi_{P_3}(f, f_1, \ldots, f_N, h_1, \ldots, h_M)_u,$$

and $\widetilde{\widetilde{g}} = \begin{cases} \phi_{P_1}(0, 0, \ldots, 0, g), & \text{on } P_1 \\ \phi_{P_2}(g, 0, \ldots, 0), & \text{on } P_2. \end{cases}$

Proof : Given maps $f, f_1, \ldots, f_N, h_1, \ldots, h_M \in C^\infty(\mathbf{S}^1, \mathbf{T}_L)$, let us put

$$f = f_* + f_0 + \lambda\theta, \quad f_i = f_{i*} + f_{i0} + \lambda_i\theta, \quad \text{and} \quad h_j = h_{j*} + h_{j0} + \nu_j\theta.$$

If $\lambda \neq \lambda_1 + \cdots + \lambda_N + \nu_1 + \cdots + \nu_M$, then both sides are equal to 0 and the statement is trivial. So let us assume that $\lambda = \lambda_1 + \cdots + \lambda_N + \nu_1 + \cdots + \nu_M$.

For each relative homotopy class u of $C^\infty{}_{P_3,L}(f, f_1, \ldots, f_N, h_1, \ldots, h_M)$, define a map $k_u \in C^\infty(\mathbf{S}^1, \mathbf{T}_L)$ by
$$k_u = \varphi_{0u}\big|_S.$$
Let us fix a relative homotopy class t' of $C^\infty{}_{P_1,L}(f, f_1, \ldots, f_N, k_u)$ and a relative homotopy class s' of $C^\infty{}_{P_2,L}(k_u, h_1, \ldots, h_M)$ such that $t' * s' = u$. Then we have
$$\varphi_{0u} = \begin{cases} \phi_{P_1}(f, f_1, \ldots, f_N, k_u)_{t'}, & \text{on } P_1, \\ \phi_{P_2}(k_u, h_1, \ldots, h_M)_{s'}, & \text{on } P_2. \end{cases}$$
Now given a map $g \in C^\infty(\mathbf{S}^1, \mathbf{T}_L)$ on S, let us put $g = g_* + g_0 + \mu\theta$. When $\mu \neq \nu_1 + \cdots + \nu_M$, then $C^\infty{}_{P_1,L}(f, f_1, \ldots, f_N, g)$ and $C^\infty{}_{P_2,L}(g, h_1, \ldots, h_M)$ are both empty. When $\mu = \nu_1 + \cdots + \nu_M$, for a relative homotopy type t of $C^\infty{}_{P_1,L}(f, f_1, \ldots, f_N, g)$ and a relative homotopy type s of $C^\infty{}_{P_2,L}(g, h_1, \ldots, h_M)$ such that $t * s = u$, define a map $\Phi_{0ts} \in C^\infty{}_{P_3,L}(f, f_1, \ldots, f_N, h_1, \ldots, h_M)_u$ by
$$\Phi_{0ts} = \begin{cases} \phi_{0t}, & \text{on } P_1, \\ \psi_{0s}, & \text{on } P_2. \end{cases}$$
Then we have
$$\Phi_{0ts} = \varphi_{0u} + \begin{cases} \phi_{P_1}(0, 0, \ldots, 0, g - k_u)_{t-t'}, & \text{on } P_1, \\ \phi_{P_2}(g - k_u, 0, \ldots, 0)_{s-s'}, & \text{on } P_2. \end{cases}$$
Let us denote the second function on the right hand side by $(\widetilde{g - k_u})_{t-t', s-s'}$. Since it has a relative homotopy type 0, we get
$$I(\Phi_{0ts}) = I(\varphi_{0u}) + I((\widetilde{g - k_u})_{t-t', s-s'}).$$
Therefore,
$$\int_{C^\infty(\mathbf{S}^1, \mathbf{T}_L)^\wedge} \sum_t e^{-I_{P_1}(\phi_{0t})} \sum_s e^{-I_{P_2}(\psi_{0s})} [dg] = \int_{C^\infty(\mathbf{S}^1, \mathbf{T}_L)^\wedge} \sum_{t,s} [e^{-I_{P_3}(\Phi_{0ts})} dg]$$
$$= \int_{C^\infty(\mathbf{S}^1, \mathbf{T}_L)^\wedge} \sum_{t,s} e^{-I_{P_3}(\varphi_{0t*s})} \cdot e^{-I_{P_3}((\widetilde{g - k_{t*s}})_{t-t', s-s'})} \cdot \delta_{\lambda+\mu, 0} [dg]$$
$$= \sum_u \int_{C^\infty(\mathbf{S}^1, \mathbf{T}_L)^\wedge} \sum_{\substack{t,s \\ t*s=u}} e^{-I_Z((\widetilde{g - k_u})_{t-t', s-s'})} \cdot \delta_{\lambda+\mu, 0} [dg] \cdot e^{-I_Z(\varphi_{0u})}$$
$$= \sum_u \int_{C^\infty(\mathbf{S}^1, \mathbf{T}_L)^\wedge} \sum_{\substack{t'',s'' \\ t''*s''=0}} e^{-I_Z((\widetilde{g - k_u})_{t'', s''})} \cdot \delta_{\lambda+\mu, 0} [dg] \cdot e^{-I_Z(\varphi_{0u})}.$$
Now for fixed u, we have
$$\int_{C^\infty(\mathbf{S}^1, \mathbf{T}_L)^\wedge} \sum_{\substack{t'',s'' \\ t''*s''=0}} e^{-I_Z((\widetilde{g - k_u})_{t'', s''})} \cdot \delta_{\lambda+\mu, 0} [dg] = \int_{C^\infty(\mathbf{S}^1, \mathfrak{h}_\mathbb{R})^\wedge} [e^{-I_Z(\widetilde{g})} dg],$$
and the proposition is proved.

80   HARUO TSUKADA

PROPOSITION **(6.9)**.

Let $P = P_{z_1,\rho_1,\ldots,z_N,\rho_N;q}$ be an N- holed disk defined in Example (5.22). Assume that $N \geq 1$. Let $\mathrm{Harm}(P, D_{z_1,\rho_1}, \ldots, D_{z_N,\rho_N})$ be the $\mathbb{R}$-vector space of functions ϕ on D_q with values in $\mathfrak{h}_{\mathbb{R}}$ such that ϕ is harmonic on $P, D_{z_1,\rho_1}, \ldots, D_{z_N,\rho_N}$ and also $\phi|_{\partial D_q} = 0$. Since any function $\phi \in \mathrm{Harm}(P, D_{z_1,\rho_1}, \ldots, D_{z_N,\rho_N})$ can be uniquely determined by its values on the circles $\partial D_{z_1,\rho_1}, \ldots, \partial D_{z_N,\rho_N}$, we have an isomorphism

$$C^\infty(\mathbf{S}^1, \mathfrak{h}_{\mathbb{R}})^N \cong \mathrm{Harm}(P, D_{z_1,\rho_1}, \ldots, D_{z_N,\rho_N}) : (g_1,\ldots,g_N) \mapsto \Psi(g_1,\ldots,g_N),$$

where

$$\Psi(g_1,\ldots,g_N) = \begin{cases} \widehat{\widetilde{g}}_1, & \text{on } D_{z_1,\rho_1}, \\ \quad \vdots \\ \widehat{\widetilde{g}}_N, & \text{on } D_{z_N,\rho_N}, \\ \phi_P(0, g_1,\ldots,g_N), & \text{on } P. \end{cases}$$

The Gaussian integral

$$\int_{(C^\infty(\mathbf{S}^1,\mathfrak{h}_{\mathbb{R}})^\wedge)^N} [e^{-I(\Psi(g_1,\ldots,g_N))} dg_1 \cdots dg_N]$$

has a finite positive value.

Proof : We prove the finiteness by induction. When $N = 1$, the statement is already proved. Let P be an $(N+1)$-holed disk. Then by Proposition (6.6), we have

$$I(\Psi(g_1,\ldots,g_N,g)) = I(\Psi(g_1,\ldots,g_N)) + I(\widetilde{g-k}).$$

Therefore if

$$\int_{(C^\infty(\mathbf{S}^1,\mathfrak{h}_{\mathbb{R}})^\wedge)^N} [e^{-I(\Psi(g_1,\ldots,g_N))} dg_1 \cdots dg_N]$$

is finite, then

$$\int_{(C^\infty(\mathbf{S}^1,\mathfrak{h}_{\mathbb{R}})^\wedge)^{N+1}} [e^{-I(\Psi(g_1,\ldots,g_N,g))} dg_1 \cdots dg_N dg]$$

is also finite.

Let us take N annuluses $A_1,\ldots,A_N \subset P$ such that the inner circle boundaries of P coincide with the inner circle boundaries of $A_1,\ldots,A_N$. We also take these annuluses disjoint. Then since the harmonic function minimizes the action I, we have

$$I(\widetilde{g}_1) + \cdots + I(\widetilde{g}_N) \geq I(\Psi(g_1,\ldots,g_N)).$$

Therefore we have

$$\prod_{i=1}^N \int_{C^\infty(\mathbf{S}^1,\mathfrak{h}_{\mathbb{R}})^\wedge} [e^{-I(\widetilde{g}_i)} dg_i] \leq \int_{(C^\infty(\mathbf{S}^1,\mathfrak{h}_{\mathbb{R}})^\wedge)^N} [e^{-I(\Psi(g_1,\ldots,g_N))} dg_1 \cdots dg_N].$$

§6-B. String Path Integrals on Disks with Holes.

DEFINITION **(6.10)**. [String Path Integrals on N-holed Disk]

Let $P = P_{z_1,\rho_1,\ldots,z_N,\rho_N;q}$ be an N-holed disk. $(N \geq 0)$ We define a constant K_P by

$$K_P = \left(\int_{(C^\infty(\mathbf{S}^1,\mathfrak{h}_\mathbb{R})^\wedge)^N} [e^{-I_{D_q}(\Psi(g_1,\ldots,g_N))} dg_1 \cdots dg_N] \right)^{-1}.$$

K_P is regarded as a string path integral

$$\int_{C^\infty{}_P(0,\ldots,0)} e^{-I_P(\phi)}[d\phi].$$

(1) For functions $f, f_1, \ldots, f_N \in C^\infty(\mathbf{S}^1, \mathfrak{h}_\mathbb{R})$, we define

$$K_P(f, f_1, \ldots, f_N) = K_P \, e^{-I_P(\phi_0)},$$

where $\phi_0 = \phi_P(f, f_1, \ldots, f_N)$. We regard it as a string path integral

$$\int_{C^\infty{}_P(f,f_1,\ldots,f_N)} e^{-I_P(\phi)}[d\phi].$$

We have

$$\int_{(C^\infty(\mathbf{S}^1,\mathfrak{h}_\mathbb{R})^\wedge)^N} K_P(0, f_1, \ldots, f_N) e^{-I(\widehat{f_1})} \cdots e^{-I(\widehat{f_N})}[df_1 \cdots df_N] = 1.$$

(2) For maps $f, f_1, \ldots, f_N \in C^\infty(\mathbf{S}^1, \mathbf{T}_L)$, we define

$$K_{P,L}(f, f_1, \ldots, f_N) = K_P \sum_{\phi_0} e^{-I_P(\phi_0)},$$

where the sum is over all $\phi_0 \in \mathrm{Harm}_{P,L}(f, f_1, \ldots, f_N)$. We regard it as a string path integral

$$\int_{C^\infty{}_{P,L}(f,f_1,\ldots,f_N)} e^{-I_P(\phi)}[d\phi].$$

Example **(6.11)**. [Disk]

Let $D = D_q$ be a disk of radius $= |q|$ defined in Example (5.20). Then we have
$$K_D = 1.$$

(1) For a function $f \in C^\infty(\mathbf{S}^1, \mathfrak{h}_\mathbb{R})$, we have
$$K_D(f) = e^{-I_D(\widehat{f})}.$$

(2) For a map $f = f_* + f_0 + \lambda\theta \in C^\infty(\mathbf{S}^1, \mathbf{T}_L)$, we have
$$K_{D,L}(f) = \delta_{\lambda,0} \cdot e^{-I_D(\widehat{f_*})}.$$

Example **(6.12)**. [Annulus]

Let $A = A_{q',q''}$ be an annulus. By Proposition (6.2), we have
$$K_A = \frac{1}{(\sqrt{2\tau_2}f(q\bar{q}))^\ell}.$$

LEMMA **(6.13)**.

For an element
$$v = L_{-1}{}^{n_1}\overline{L}_{-1}{}^{m_1}e^{(\beta_1,\beta_1)}\cdots L_{-1}{}^{n_N}\overline{L}_{-1}{}^{m_N}e^{(\beta_N,\beta_N)},$$
of $U_{L'}$, define a differential operator
$$D = \left(\frac{\partial}{\partial z_1}\right)^{n_1}\left(\frac{\partial}{\partial \bar{z}_1}\right)^{m_1}\cdots\left(\frac{\partial}{\partial z_N}\right)^{n_N}\left(\frac{\partial}{\partial \bar{z}_N}\right)^{m_N}\Bigg|_{z_1=\cdots=z_N=0},$$
then we have the following.

(1) The value $v(g)$ for $g = g_* + g_0 + \mu\theta \in C^\infty(\mathbf{S}^1, \mathbf{T}_L)$ is given by
$$v(g) = \delta_{\mu,0} \cdot D\left\{{}_\bullet^\bullet e^{i<\beta_1,\widehat{g}(z_1)>}\cdots e^{i<\beta_N,\widehat{g}(z_N)>}{}_\bullet^\bullet\right\},$$
where $\widehat{g} = \phi_D(g)$ on the unit disk $D = D_1$.

(2) The value $(q^{-L_0}\bar{q}^{-\overline{L}_0} \cdot v)(g)$ for $g = g_* + g_0 + \mu\theta \in C^\infty(\mathbf{S}^1, \mathbf{T}_L)$ is given by
$$(q^{-L_0}\bar{q}^{-\overline{L}_0} \cdot v)(g) = |q|^{-<\beta,\beta>} \cdot \delta_{\mu,0} \cdot D\left\{{}_\bullet^\bullet e^{i<\beta_1,\widehat{g}(z_1)>}\cdots e^{i<\beta_N,\widehat{g}(z_N)>}{}_\bullet^\bullet\right\},$$
where $\widehat{g} = \phi_{D_q}(g)$ on the disk D_q of radius $= |q|$, and $\beta = \beta_1 + \cdots + \beta_N$.

THEOREM (6.14). [Annulus and the Degree Operators $L_0, \overline{L}_0$]

Let $\mathcal{K}$ be the Hilbert completion of $U_{L'}$. Let $A = A_{q',q''}$ be an annulus. The functional $K_{A,L}(f,g)$ is the integral kernel of an operator $\mathcal{K} \to \mathcal{K}$ which is the extension of the operator

$$q^{L_0}\overline{q}^{\overline{L}_0} : U_{L'} \to U_{L'},$$

where $q = q''/q'$. Namely we have

$$\int_{C^\infty(S^1,\mathbf{T}_L)^\wedge} K_{A,L}(f,g)v(g)\, e^{-\frac{1}{2}(g_*,\overline{g}_*)}[dg] = \left(q^{L_0}\overline{q}^{\overline{L}_0}\cdot v\right)(f)\, e^{-\frac{1}{2}(f_*,\overline{f}_*)},$$

for $v \in U_{L'}$.

Proof: We will show that

$$\int_{C^\infty(S^1,\mathbf{T}_L)^\wedge} K_{A,L}(f,g)\left({q''}^{-L_0}\overline{q''}^{-\overline{L}_0}\cdot v\right)(g)\, e^{-\frac{1}{2}(g_*,\overline{g}_*)}[dg] = \left({q'}^{-L_0}\overline{q'}^{-\overline{L}_0}\cdot v\right)(f)\, e^{-\frac{1}{2}(f_*,\overline{f}_*)}.$$

Because of Corollary (1.12), we can assume that

$$v = L_{-1}{}^{n_1}\overline{L}_{-1}{}^{m_1}e^{(\beta_1,\beta_1)}\cdots L_{-1}{}^{n_N}\overline{L}_{-1}{}^{m_N}e^{(\beta_N,\beta_N)},$$

without loss of generality. We have

$$\int_{C^\infty(S^1,\mathbf{T}_L)^\wedge} K_{A,L}(f,g)\cdot|q''|^{-<\beta,\beta>}\cdot\delta_{\mu,0}\cdot{}^\bullet_\bullet e^{i<\beta_1,\widehat{g}(z_1)>}\cdots e^{i<\beta_N,\widehat{g}(z_N)>}{}^\bullet_\bullet e^{-I(\widehat{g})}[dg]$$

$$= K_A\sum_t \int_{C^\infty(S^1,\mathbf{T}_L)^\wedge} e^{-I_A(\phi_A(f,g)_t)}\cdot|q''|^{-<\beta,\beta>}$$
$$\cdot\delta_{\mu,0}\cdot{}^\bullet_\bullet e^{i<\beta_1,\widehat{g}(z_1)>}\cdots e^{i<\beta_N,\widehat{g}(z_N)>}{}^\bullet_\bullet e^{-I(\widehat{g})}[dg]$$

$$= K_A\cdot|q''|^{-<\beta,\beta>}\cdot\delta_{\lambda,0}\cdot e^{i<\beta_1,\widehat{f}(z_1)>}\cdots e^{i<\beta_N,\widehat{f}(z_N)>}e^{-I(\widehat{f})}$$
$$\cdot\int_{C^\infty(S^1,\mathfrak{h}_\mathbf{R})^\wedge} {}^\bullet_\bullet e^{i<\beta_1,\widehat{g}(z_1)>}\cdots e^{i<\beta_N,\widehat{g}(z_N)>}{}^\bullet_\bullet[e^{-I(\phi_A(0,g))-I(\widehat{g})}dg]$$

$$= K_A\cdot|q''|^{-<\beta,\beta>}\cdot\delta_{\lambda,0}\cdot e^{i<\beta_1,\widehat{f}(z_1)>}\cdots e^{i<\beta_N,\widehat{f}(z_N)>}e^{-I(\widehat{f})}$$
$$\cdot\prod_{i,j}\left|\frac{1}{1-z_i''\overline{z_j''}}\right|^{<\beta_i,\beta_j>}\cdot\int_{C^\infty(S^1,\mathfrak{h}_\mathbf{R})^\wedge} e^{i<\beta_1,\widehat{g}(z_1)>}\cdots e^{i<\beta_N,\widehat{g}(z_N)>}[e^{-I(\phi_A(0,g))-I(\widehat{g})}dg]$$

$$= |q''|^{-<\beta,\beta>}\cdot\delta_{\lambda,0}\cdot e^{i<\beta_1,\widehat{f}(z_1)>}\cdots e^{i<\beta_N,\widehat{f}(z_N)>}e^{-I(\widehat{f})}$$
$$\cdot\prod_{i,j}\left\{\left|\frac{1}{1-z_i''\overline{z_j''}}\right|\cdot\left|\frac{z_i'-z_j'}{1-z_i'\overline{z_j'}}\right|\cdot\left|\frac{1-z_i''\overline{z_j''}}{z_i''-z_j''}\right|\right\}^{<\beta_i,\beta_j>}$$

$$= |q'|^{-<\beta,\beta>}\cdot\delta_{\lambda,0}\cdot{}^\bullet_\bullet e^{i<\beta_1,\widehat{f}(z_1)>}\cdots e^{i<\beta_N,\widehat{f}(z_N)>}{}^\bullet_\bullet e^{-I(\widehat{f})},$$

where $\widehat{f} = \phi_{D_{q'}}(f), \widehat{g} = \phi_{D_{q''}}(g)$, and $z_i' = z_i/q', z_i'' = z_i/q''$. Differentiating the both sides by

$$\left(\frac{\partial}{\partial z_1}\right)^{n_1} \left(\frac{\partial}{\partial \overline{z}_1}\right)^{m_1} \cdots \left(\frac{\partial}{\partial z_N}\right)^{n_N} \left(\frac{\partial}{\partial \overline{z}_N}\right)^{m_N}\bigg|_{z_1 = \cdots = z_N = 0},$$

and using Lemma (6.13), the desired equation follows.

Lemma **(6.15).**

We have the following functional realization of the products of neutral vertex operators.

(1) $\left(\begin{smallmatrix}\circ\\\circ\end{smallmatrix}Y(e^{(\alpha_1,\alpha_1)}, z_1) \cdots Y(e^{(\alpha_N,\alpha_N)}, z_N)\begin{smallmatrix}\circ\\\circ\end{smallmatrix} \cdot e^{(0,0)}\right)(f) = \delta_{\lambda,0} \cdot {}^\bullet_\bullet e^{i<\alpha_1,\widehat{f}(z_1)>} \cdots e^{i<\alpha_N,\widehat{f}(z_N)>}{}^\bullet_\bullet,$

(2) $\left(Y(e^{(\alpha_1,\alpha_1)}, z_1) \cdots Y(e^{(\alpha_N,\alpha_N)}, z_N) \cdot e^{(0,0)}\right)(f)$

$$= \prod_{i \neq j} \left|\frac{z_i - z_j}{1 - z_i\overline{z}_j}\right|^{<\alpha_i,\alpha_j>} \cdot \delta_{\lambda,0} \cdot {}^\bullet_\bullet e^{i<\alpha_1,\widehat{f}(z_1)>}{}^\bullet_\bullet \cdots {}^\bullet_\bullet e^{i<\alpha_N,\widehat{f}(z_N)>}{}^\bullet_\bullet.$$

where $\widehat{f} = \phi_D(f)$ on the unit disk $D = D_1$. Notice the appearance of the Green function on the unit disk.

Lemma **(6.16).**

For an element

$$v = L_{-1}{}^{n_1}\overline{L}_{-1}{}^{m_1} e^{(\beta_1,\beta_1)} \cdots L_{-1}{}^{n_N}\overline{L}_{-1}{}^{m_N} e^{(\beta_N,\beta_N)},$$

of $U_{L'}$, define a differential operator

$$D = \left(\frac{\partial}{\partial z_1}\right)^{n_1} \left(\frac{\partial}{\partial \overline{z}_1}\right)^{m_1} \cdots \left(\frac{\partial}{\partial z_N}\right)^{n_N} \left(\frac{\partial}{\partial \overline{z}_N}\right)^{m_N}\bigg|_{z_1 = \cdots = z_N = 0},$$

then we have the following.

(1) For $|z| < 1$, the value $(\exp(-zL_{-1})\exp(-\overline{z}\overline{L}_{-1}) \cdot v)(g)$ for $g = g_* + g_0 + \mu\theta \in C^\infty(\mathbf{S}^1, \mathbf{T}_L)$ is given by

$$\left(\exp(-zL_{-1})\exp(-\overline{z}\overline{L}_{-1}) \cdot v\right)(g) = \delta_{\mu,0} \cdot D\left\{{}^\bullet_\bullet e^{i<\beta_1,\widehat{\widehat{g}}(z_1)>} \cdots e^{i<\beta_N,\widehat{\widehat{g}}(z_N)>}{}^\bullet_\bullet\right\},$$

where $\widehat{\widehat{g}} = \phi_{D_{z,1}}(g)$ on the unit disk $D_{z,1}$ centered at z.

(2) For $|z| < \rho$, the value $(\rho^{-L_0}\overline{\rho}^{-\overline{L}_0}\exp(-zL_{-1})\exp(-\overline{z}\overline{L}_{-1})\cdot v)(g)$ for $g = g_*+g_0+\mu\theta \in C^\infty(\mathbf{S}^1, \mathbf{T}_L)$ is given by

$$\left(\rho^{-L_0}\overline{\rho}^{-\overline{L}_0}\exp(-zL_{-1})\exp(-\overline{z}\overline{L}_{-1}) \cdot v\right)(g)$$

$$= |\rho|^{-<\beta,\beta>} \cdot \delta_{\mu,0} \cdot D\left\{{}^\bullet_\bullet e^{i<\beta_1,\widehat{\widehat{g}}(z_1)>} \cdots e^{i<\beta_N,\widehat{\widehat{g}}(z_N)>}{}^\bullet_\bullet\right\},$$

where $\widehat{\widehat{g}} = \phi_{D_{z,\rho}}(g)$ on the disk $D_{z,\rho}$ of radius $= |\rho|$ centered at z, and $\beta = \beta_1 + \cdots + \beta_N$.

Proof : By Proposition (2.24), we have

$$\big(\exp(-zL_{-1})\exp(-\overline{z}\overline{L}_{-1}) \cdot v\big)(g)$$

$$= \big(Y(u,-z) \cdot e^{(0,0)}\big)(g) = D\big(\,{}_\circ^\circ Y(e^{(\beta_1,\beta_1)},-z_1)\cdots Y(e^{(\beta_N,\beta_N)},-z_N)_\circ^\circ \cdot e^{(0,0)}\big)(g).$$

Now using Lemma (6.15), we get

$$= \delta_{\mu,0} \cdot D\big(\,{}_\bullet^\bullet e^{i<\beta_1,\widehat{g}(z_1-z)>} \cdots e^{i<\beta_N,\widehat{g}(z_N-z)>}{}_\bullet^\bullet\big),$$

where $\widehat{g} = \phi_D(g)$. Since $\widehat{g}(-z) = \widehat{\overline{g}}(0)$, we get (1) Combining this with Lemma (6.13), we get (2).

THEOREM **(6.17)**. [1-holed Disk and the Virasoro Operators $L_0, \overline{L}_0, L_{-1}, \overline{L}_{-1}$]
Let $U = P_{z,\rho;q'}$ be a 1-holed disk. (See Example (5.22).)

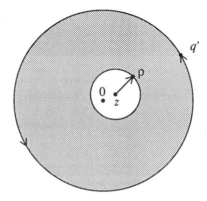

The functional $K_{U,L}(f,g)$ is the integral kernel of an operator $\mathcal{K} \to \mathcal{K}$ which is the extension of the operator

$$q'^{-L_0}\overline{q'}^{-\overline{L}_0}\exp(zL_{-1})\exp(\overline{z}\overline{L}_{-1})\rho^{L_0}\overline{\rho}^{\overline{L}_0} : U_{L'} \to \mathcal{K}.$$

Namely we have

$$\int_{C^\infty(\mathbf{S}^1,\mathbf{T}_L)^\wedge} K_{U,L}(f,g)v(g) \, e^{-\frac{1}{2}(g_*,\overline{g}_*)}[dg]$$

$$= \Big(q'^{-L_0}\overline{q'}^{-\overline{L}_0}\exp(zL_{-1})\exp(\overline{z}\overline{L}_{-1})\rho^{L_0}\overline{\rho}^{\overline{L}_0} \cdot v\Big)(f) \, e^{-\frac{1}{2}(f_*,\overline{f}_*)},$$

for $v \in U_{L'}$.

Proof : We will show that

$$\int_{C^\infty(\mathbf{S}^1,\mathbf{T}_L)^\wedge} K_{U,L}(f,g)\Big(\rho^{L_0}\overline{\rho}^{\overline{L}_0}\exp(-zL_{-1})\exp(-\overline{z}\overline{L}_{-1}) \cdot v\Big)(g) \, e^{-\frac{1}{2}(g_*,\overline{g}_*)}[dg]$$

$$= \Big(q'^{-L_0}\overline{q'}^{-\overline{L}_0} \cdot v\Big)(f) \, e^{-\frac{1}{2}(f_*,\overline{f}_*)}.$$

Because of Corollary (1.12), we can assume that

$$v = L_{-1}{}^{n_1} \overline{L}_{-1}{}^{m_1} e^{(\beta_1, \beta_1)} \cdots L_{-1}{}^{n_N} \overline{L}_{-1}{}^{m_N} e^{(\beta_N, \beta_N)},$$

without loss of generality. We have

$$\int_{C^\infty(\mathbf{S}^1, \mathbf{T}_L)^\wedge} K_{U,L}(f,g) \cdot |\rho|^{-<\beta,\beta>} \cdot \delta_{\mu,0} \cdot {}^\bullet_\bullet e^{i<\beta_1, \widehat{\widehat{g}}(z_1)>} \ldots e^{i<\beta_N, \widehat{\widehat{g}}(z_N)>}{}^\bullet_\bullet e^{-I(\widehat{\widehat{g}})}[dg]$$

$$= K_U \sum_t \int_{C^\infty(\mathbf{S}^1, \mathbf{T}_L)^\wedge} e^{-I_U(\phi_U(f,g)_t)} \cdot |\rho|^{-<\beta,\beta>}$$

$$\cdot \delta_{\mu,0} \cdot {}^\bullet_\bullet e^{i<\beta_1, \widehat{\widehat{g}}(z_1)>} \ldots e^{i<\beta_N, \widehat{\widehat{g}}(z_N)>}{}^\bullet_\bullet e^{-I(\widehat{\widehat{g}})}[dg]$$

$$= K_U \cdot |\rho|^{-<\beta,\beta>} \cdot \delta_{\lambda,0} \cdot e^{i<\beta_1, \widehat{f}(z_1)>} \ldots e^{i<\beta_N, \widehat{f}(z_N)>} e^{-I(\widehat{f})}$$

$$\cdot \int_{C^\infty(\mathbf{S}^1, \mathfrak{h}_{\mathbf{R}})^\wedge} {}^\bullet_\bullet e^{i<\beta_1, \widehat{\widehat{g}}(z_1)>} \ldots e^{i<\beta_N, \widehat{\widehat{g}}(z_N)>}{}^\bullet_\bullet [e^{-I(\phi_U(0,g)) - I(\widehat{\widehat{g}})} dg]$$

$$= K_U \cdot |\rho|^{-<\beta,\beta>} \cdot \delta_{\lambda,0} \cdot e^{i<\beta_1, \widehat{f}(z_1)>} \ldots e^{i<\beta_N, \widehat{f}(z_N)>} e^{-I(\widehat{f})} \cdot \prod_{i,j} \left| \frac{1}{1 - \frac{z_i - z}{\rho} \overline{\frac{z_j - z}{\rho}}} \right|^{<\beta_i, \beta_j>}$$

$$\cdot \int_{C^\infty(\mathbf{S}^1, \mathfrak{h}_{\mathbf{R}})^\wedge} e^{i<\beta_1, \widehat{\widehat{g}}(z_1)>} \ldots e^{i<\beta_N, \widehat{\widehat{g}}(z_N)>} [e^{-I(\phi_U(0,g)) - I(\widehat{\widehat{g}})} dg]$$

$$= |\rho|^{-<\beta,\beta>} \cdot \delta_{\lambda,0} \cdot e^{i<\beta_1, \widehat{f}(z_1)>} \ldots e^{i<\beta_N, \widehat{f}(z_N)>} e^{-I(\widehat{f})}$$

$$\prod_{i,j} \left\{ \left| \frac{1}{1 - \frac{z_i - z}{\rho} \overline{\frac{z_j - z}{\rho}}} \right| \cdot \left| \frac{z_i' - z_j'}{1 - z_i' \overline{z_j'}} \right| \cdot \left| \frac{1 - \frac{z_i - z}{\rho} \overline{\frac{z_j - z}{\rho}}}{\frac{z_i - z}{\rho} - \frac{z_j - z}{\rho}} \right| \right\}^{<\beta_i, \beta_j>}$$

$$= |q'|^{-<\beta,\beta>} \cdot \delta_{\lambda,0} \cdot {}^\bullet_\bullet e^{i<\beta_1, \widehat{f}(z_1)>} \ldots e^{i<\beta_N, \widehat{f}(z_N)>}{}^\bullet_\bullet e^{-I(\widehat{f})},$$

where $\widehat{f} = \phi_{D_{q'}}(f), \widehat{\widehat{g}} = \phi_{D_{z,\rho}}(g)$, and $z_i' = z_i/q'$. Differentiating the both sides by

$$\left(\frac{\partial}{\partial z_1} \right)^{n_1} \left(\frac{\partial}{\partial \overline{z}_1} \right)^{m_1} \cdots \left(\frac{\partial}{\partial z_N} \right)^{n_N} \left(\frac{\partial}{\partial \overline{z}_N} \right)^{m_N} \Big|_{z_1 = \cdots = z_N = 0},$$

and using Lemma (6.16), we get the equation we wanted.

§6-C. Geometric Realization of Vertex Operators.

In this section, we show that the neutral vertex operators can be realized by the string path integral over 2-holed disk. By sewing two of them, we get the geometric proof of the associativity of the neutral vertex operators.

LEMMA (6.18).

We have the following geometric realization

$$\left({}_\circ^\circ Y(e^{(\alpha_1,\alpha_1)}, z_1) \cdots Y(e^{(\alpha_N,\alpha_N)}, z_N)_\circ^\circ \cdot {}_\circ^\circ Y(e^{(\beta_1,\beta_1)}, w_1) \cdots Y(e^{(\beta_N,\beta_N)}, w_N)_\circ^\circ \cdot e^{(0,0)} \right)(f)$$

$$= \prod_{i,j} \left| \frac{z_i - w_j}{1 - z_i \overline{w_j}} \right|^{2 <\alpha_i, \beta_j>} \cdot \delta_{\lambda,0} \cdot {}_\bullet^\bullet e^{i<\alpha_1, \widehat{f}(z_1)>} \cdots e^{i<\alpha_N, \widehat{f}(z_N)>} {}_\bullet^\bullet$$

$$\qquad\qquad\qquad\qquad {}_\bullet^\bullet e^{i<\beta_1, \widehat{f}(w_1)>} \cdots e^{i<\beta_N, \widehat{f}(w_M)>} {}_\bullet^\bullet.$$

(See also Lemma (6.15).)

THEOREM (6.19). [Pants and Vertex Operator]

Let $P = P_{0,q'',z,\rho;q'}$ be a 2-holed disk (pants).

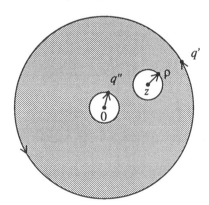

Then the functional $K_{P,L}(f,g,h)$ is the integral kernel of an operator $\mathcal{K} \otimes \mathcal{K} \to \mathcal{K}$ which is the extension of the operator

$$q'^{-L_0} \overline{q'}^{-\overline{L}_0} Y(\rho^{L_0} \overline{\rho}^{\overline{L}_0}(\ ), z) q''^{L_0} \overline{q''}^{\overline{L}_0} : U_{L'} \otimes U_{L'} \to \mathcal{K}.$$

Namely we have

$$\int_{Cm(S^1,T_*)^\wedge} \int_{Cm(S^1,T_*)^\wedge} K_{P,L}(f,g,h) u(g)\, e^{-\frac{1}{2}(g_*,\overline{g}_*)} v(h)\, e^{-\frac{1}{2}(h_*,\overline{h}_*)} [dgdh]$$

$$= \left(q'^{-L_0} \overline{q'}^{-\overline{L}_0} Y(\rho^{L_0} \overline{\rho}^{\overline{L}_0} u, z) q''^{L_0} \overline{q''}^{\overline{L}_0} \cdot v \right)(f)\, e^{-\frac{1}{2}(f_*,\overline{f}_*)},$$

for all u, $v \in U_{L'}$.

Remark. This theorem justifies our use of the letter "Y" for the vertex operators, since a Y-shaped pipe is conformally equivalent to a 2-holed disk. (This was originally pointed out by Professor James Lepowsky.)

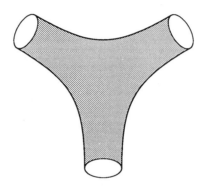

Proof : We will show that

$$\int_{C^\infty(\mathbf{S}^1,\mathbf{T}_L)^\wedge}\int_{C^\infty(\mathbf{S}^1,\mathbf{T}_L)^\wedge} K_{P,L}(f,g,h)\left(\rho^{-L_0}\overline{\rho}^{-\overline{L}_0}\cdot u\right)(g)\,e^{-\frac{1}{2}(g_*,\overline{g}_*)}$$
$$\cdot\left(q''^{-L_0}\overline{q''}^{-\overline{L}_0}\cdot v\right)(h)\,e^{-\frac{1}{2}(h_*,\overline{h}_*)}[dg\,dh]$$

$$=\left(q'^{-L_0}\overline{q'}^{-\overline{L}_0}Y(u,z)\cdot v\right)(f)\,e^{-\frac{1}{2}(f_*,\overline{f}_*)}.$$

Because of Corollary (1.12), we can assume that

$$u=L_{-1}{}^{n_1}\overline{L}_{-1}{}^{m_1}e^{(\alpha_1,\alpha_1)}\cdots L_{-1}{}^{n_N}\overline{L}_{-1}{}^{m_N}e^{(\alpha_N,\alpha_N)},$$
$$v=L_{-1}{}^{k_1}\overline{L}_{-1}{}^{l_1}e^{(\beta_1,\beta_1)}\cdots L_{-1}{}^{k_M}\overline{L}_{-1}{}^{l_M}e^{(\beta_M,\beta_M)},$$

without loss of generality. We have

$$\int_{C^\infty(\mathbf{S}^1,\mathbf{T}_L)^\wedge}\int_{C^\infty(\mathbf{S}^1,\mathbf{T}_L)^\wedge} K_{P,L}(f,g,h)$$
$$\cdot|\rho|^{-<\alpha,\alpha>}\cdot\delta_{\mu,0}\cdot{}^\bullet_\bullet e^{i<\alpha_1,\widehat{g}(z_1)>}\ldots e^{i<\alpha_N,\widehat{g}(z_N)>}{}^\bullet_\bullet e^{-I(\widehat{g})}$$
$$\cdot|q''|^{-<\beta,\beta>}\cdot\delta_{\nu,0}\cdot{}^\bullet_\bullet e^{i<\beta_1,\widehat{h}(w_1)>}\ldots e^{i<\beta_M,\widehat{h}(w_M)>}{}^\bullet_\bullet e^{-I(\widehat{h})}[dg\,dh]$$

$$=K_P\sum_t\int_{C^\infty(\mathbf{S}^1,\mathbf{T}_L)^\wedge}\int_{C^\infty(\mathbf{S}^1,\mathbf{T}_L)^\wedge}e^{-I_P(\phi_P(f,g,h)_t)}$$
$$\cdot|\rho|^{-<\alpha,\alpha>}\cdot\delta_{\mu,0}\cdot{}^\bullet_\bullet e^{i<\alpha_1,\widehat{g}(z_1)>}\ldots e^{i<\alpha_N,\widehat{g}(z_N)>}{}^\bullet_\bullet e^{-I(\widehat{g})}$$
$$\cdot|q''|^{-<\beta,\beta>}\cdot\delta_{\nu,0}\cdot{}^\bullet_\bullet e^{i<\beta_1,\widehat{h}(w_1)>}\ldots e^{i<\beta_M,\widehat{h}(w_M)>}{}^\bullet_\bullet e^{-I(\widehat{h})}[dg\,dh]$$

$$= K_P \cdot |\rho|^{-<\alpha,\alpha>} \cdot |q''|^{-<\beta,\beta>} \cdot \delta_{\lambda,0}$$

$$\cdot e^{i<\alpha_1,\widehat{f}(z_1+z)>} \dots e^{i<\alpha_N,\widehat{f}(z_N+z)>} \cdot e^{i<\beta_1,\widehat{f}(w_1)>} \dots e^{i<\beta_M,\widehat{f}(w_M)>} e^{-I(\widehat{f})}$$

$$\cdot \int_{C^\infty(\mathbf{S}^1,\mathfrak{h}_{\mathbf{R}})^\wedge} \int_{C^\infty(\mathbf{S}^1,\mathfrak{h}_{\mathbf{R}})^\wedge} {}^\bullet_\bullet e^{i<\alpha_1,\widehat{g}(z_1)>} \dots e^{i<\alpha_N,\widehat{g}(z_N)>} {}^\bullet_\bullet$$

$$\cdot {}^\bullet_\bullet e^{i<\beta_1,\widehat{h}(w_1)>} \dots e^{i<\beta_M,\widehat{h}(w_M)>} {}^\bullet_\bullet [e^{-I_P(\phi_P(0,g,h))-I(\widehat{g})-I(\widehat{h})} dg\, dh]$$

$$= K_P \cdot |\rho|^{-<\alpha,\alpha>} \cdot |q''|^{-<\beta,\beta>} \cdot \delta_{\lambda,0}$$

$$\cdot e^{i<\alpha_1,\widehat{f}(z_1+z)>} \dots e^{i<\alpha_N,\widehat{f}(z_N+z)>} \cdot e^{i<\beta_1,\widehat{f}(w_1)>} \dots e^{i<\beta_M,\widehat{f}(w_M)>} e^{-I(\widehat{f})}$$

$$\cdot \prod_{i,j} \left| \frac{1}{1-\frac{z_i}{\rho}\frac{\overline{z_j}}{\overline{\rho}}} \right|^{<\alpha_i,\alpha_j>} \cdot \prod_{i,j} \left| \frac{1}{1-w_i''\overline{w_j''}} \right|^{<\beta_i,\beta_j>}$$

$$\cdot \int_{C^\infty(\mathbf{S}^1,\mathfrak{h}_{\mathbf{R}})^\wedge} \int_{C^\infty(\mathbf{S}^1,\mathfrak{h}_{\mathbf{R}})^\wedge} e^{i<\alpha_1,\widehat{g}(z_1)>} \dots e^{i<\alpha_N,\widehat{g}(z_N)>}$$

$$\cdot e^{i<\beta_1,\widehat{h}(w_1)>} \dots e^{i<\beta_M,\widehat{h}(w_M)>} [e^{-I_P(\phi_P(0,g,h))-I(\widehat{g})-I(\widehat{h})} dg\, dh]$$

$$= K_P \cdot |\rho|^{-<\alpha,\alpha>} \cdot |q''|^{-<\beta,\beta>} \cdot \delta_{\lambda,0}$$

$$\cdot e^{i<\alpha_1,\widehat{f}(z_1+z)>} \dots e^{i<\alpha_N,\widehat{f}(z_N+z)>} \cdot e^{i<\beta_1,\widehat{f}(w_1)>} \dots e^{i<\beta_M,\widehat{f}(w_M)>} e^{-I(\widehat{f})}$$

$$\cdot \prod_{i,j} \left\{ \left| \frac{1}{1-\frac{z_i}{\rho}\frac{\overline{z_j}}{\overline{\rho}}} \right| \cdot \left| \frac{\frac{z_i+z}{q'} - \frac{z_j+z}{q'}}{1-\frac{z_i+z}{q'}\overline{\frac{z_j+z}{q'}}} \right| \cdot \left| \frac{1-\frac{z_i}{\rho}\frac{\overline{z_j}}{\overline{\rho}}}{\frac{z_i}{\rho} - \frac{z_j}{\rho}} \right| \right\}^{<\alpha_i,\alpha_j>}$$

$$\cdot \prod_{i,j} \left\{ \left| \frac{1}{1-w_i''\overline{w_j''}} \right| \cdot \left| \frac{w_i' - w_j'}{1-w_i'\overline{w_j'}} \right| \cdot \left| \frac{1-w_i''\overline{w_j''}}{w_i'' - w_j''} \right| \right\}^{<\beta_i,\beta_j>} \cdot \prod_{i,j} \left| \frac{\frac{z_i+z}{q'} - w_j'}{1-\frac{z_i+z}{q'}\overline{w_j'}} \right|^{2<\alpha_i,\beta_j>}$$

where $w_i' = w_i/q'$, $w_i'' = w_i/q''$, $\widehat{f} = \phi_{D_{q'}}(f)$, $\widehat{g} = \phi_{D_\rho}(g)$, $\widehat{h} = \phi_{D_{q''}}(h)$, and $\alpha = \alpha_1 + \dots + \alpha_N$, $\beta = \beta_1 + \dots + \beta_M$. By using Lemma (6.18), we get

$$= K_P \cdot |q'|^{-<\alpha+\beta,\alpha+\beta>} \cdot \delta_{\lambda,0}$$

$$\cdot {}^\bullet_\bullet e^{i<\alpha_1,\widehat{f}(z_1+z)>} \dots e^{i<\alpha_N,\widehat{f}(z_N+z)>} {}^\bullet_\bullet \cdot {}^\bullet_\bullet e^{i<\beta_1,\widehat{f}(w_1)>} \dots e^{i<\beta_M,\widehat{f}(w_M)>} {}^\bullet_\bullet e^{-I(\widehat{f})}$$

$$\cdot \prod_{i,j} \left| \frac{(z_i+z) - w_j}{1-\frac{z_i+z}{q'}\overline{w_j'}} \right|^{2<\alpha_i,\beta_j>}$$

$$= q'^{-L_0} \overline{q'}^{-\overline{L_0}} \left\{ {}^\circ_\circ Y(e^{(\alpha_1,\alpha_1)}, z_1+z) \dots Y(e^{(\alpha_N,\alpha_N)}, z_N+z) {}^\circ_\circ \right.$$

$$\left. {}^\circ_\circ Y(e^{(\beta_1,\beta_1)}, w_1) \dots Y(e^{(\beta_M,\beta_M)}, w_M) {}^\circ_\circ \cdot e^{(0,0)} \right\}(f) \, e^{-I(\widehat{f})}.$$

Differentiating the both sides by

$$\left(\frac{\partial}{\partial z_1} \right)^{n_1} \left(\frac{\partial}{\partial \overline{z_1}} \right)^{m_1} \dots \left(\frac{\partial}{\partial z_N} \right)^{n_N} \left(\frac{\partial}{\partial \overline{z_N}} \right)^{m_N} \bigg|_{z_1 = \dots = z_N = 0},$$

and by

$$\left(\frac{\partial}{\partial w_1}\right)^{k_1}\left(\frac{\partial}{\partial \overline{w}_1}\right)^{l_1}\cdots\left(\frac{\partial}{\partial w_M}\right)^{k_M}\left(\frac{\partial}{\partial \overline{w}_M}\right)^{l_M}\Bigg|_{w_1=\cdots=w_M=0},$$

and using Lemma (6.13), we get the equation we wanted.

THEOREM (6.20). [Sewing of Two Disks with Holes]
Let $P_1 = P_{z_1,\rho_1,\ldots,z_N,\rho_N,z,\rho;q}$ be an $N+1$-holed disk and $P_2 = P_{w_1-z,r_1,\ldots,w_M-z,r_M;\rho}$ be an M-holed disk. Let $P_3 = P_{z_1,\rho_1,\ldots,z_N,\rho_N,w_1,r_1,\ldots,w_M,r_M;q}$ be the $N+M$-holed disk which is obtained by sewing P_1 and P_2. Then we have

(1) $\displaystyle\int_{C^\infty(\mathbf{S}^1,\mathfrak{h}_{\mathbb{R}})^\wedge} K_{P_1}(f,f_1,\ldots,f_N,g)\,K_{P_2}(g,h_1,\ldots,h_M)[dg]$

$$= K_{P_3}(f,f_1,\ldots,f_N,h_1,\ldots,h_M).$$

(2) $\displaystyle\int_{C^\infty(\mathbf{S}^1,\mathbf{T}_L)^\wedge} K_{P_1,L}(f,f_1,\ldots,f_N,g)\,K_{P_2,L}(g,h_1,\ldots,h_M)[dg]$

$$= K_{P_3,L}(f,f_1,\ldots,f_N,h_1,\ldots,h_M).$$

Proof : First, we prove the equality

$$K_{P_3} = K_{P_1}K_{P_2}\int_{C^\infty(\mathbf{S}^1,\mathfrak{h}_{\mathbb{R}})^\wedge}[e^{-I_{P_3}(\widetilde{\widetilde{g}})}dg],$$

which is equivalent to

$$K_{P_1}K_{P_2}\int_{(C^\infty(\mathbf{S}^1,\mathfrak{h}_{\mathbb{R}})^\wedge)^{N+M+1}}[e^{-I_{P_3}(\widetilde{\widetilde{g}})}\cdot e^{-I_{P_3}(\phi_{P_3}(0,f_1,\ldots,f_N,h_1,\ldots,h_M))}$$

$$\cdot e^{-I_D(\widehat{f}_1)}\ldots e^{-I_D(\widehat{f}_N)}\cdot e^{-I_D(\widehat{h}_1)}\ldots e^{-I_D(\widehat{h}_M)}df_1\cdots df_N dg dh_1\cdots dh_M] = 1.$$

By Proposition (6.6) for sewing of P_1 and P_2, the left hand side is equal to

$$K_{P_1}K_{P_2}\int_{(C^\infty(\mathbf{S}^1,\mathfrak{h}_{\mathbb{R}})^\wedge)^{N+M+1}}[e^{-I_{P_1}(\phi_{P_1}(0,f_1,\ldots,f_N,g))}\cdot e^{-I_{P_2}(\phi_{P_2}(g,h_1,\ldots,h_M))}$$

$$\cdot e^{-I_D(\widehat{f}_1)}\ldots e^{-I_D(\widehat{f}_N)}\cdot e^{-I_D(\widehat{h}_1)}\ldots e^{-I_D(\widehat{h}_M)}df_1\cdots df_N dg dh_1\cdots dh_M].$$

Using Proposition (6.6) again this time for sewing of P_2 and N disks, we get

$$= K_{P_1}K_{P_2}$$

$$\cdot\int_{(C^\infty(\mathbf{S}^1,\mathfrak{h}_{\mathbb{R}})^\wedge)^{N+1}}[e^{-I_{P_1}(\phi_{P_1}(0,f_1,\ldots,f_N,g))}\cdot e^{-I_D(\widehat{g})}\cdot e^{-I_D(\widehat{f}_1)}\ldots e^{-I_D(\widehat{f}_N)}df_1\cdots df_N dg]$$

$$\cdot\int_{(C^\infty(\mathbf{S}^1,\mathfrak{h}_{\mathbb{R}})^\wedge)^M}[e^{-I_{P_2}(\phi_{P_2}(0,h_1,\ldots,h_M))}\cdot e^{-I_D(\widehat{h}_1)}\ldots e^{-I_D(\widehat{h}_M)}dh_1\cdots dh_M].$$

By the definition of K_{P_2}, we have

$$= K_{P_1}$$
$$\cdot \int_{(C^\infty(\mathbf{S}^1, \mathfrak{h}_{\mathbf{R}})^\wedge)^{N+1}} [e^{-I_{P_1}(\phi_{P_1}(0, f_1, \ldots, f_N, g))} \cdot e^{-I_D(\widehat{g})} \cdot e^{-I_D(\widehat{f_1})} \ldots e^{-I_D(\widehat{f_N})} df_1 \cdots df_N dg].$$

Again by the definition of K_{P_1}, we see that the above value is equal to 1.

(1) Now by Proposition (6.6), we have

$$\int_{C^\infty(\mathbf{S}^1, \mathfrak{h}_{\mathbf{R}})^\wedge} K_{P_1}(f, f_1, \ldots, f_N, g) \, K_{P_2}(g, h_1, \ldots, h_M)[dg]$$

$$= K_{P_1} K_{P_2} \int_{C^\infty(\mathbf{S}^1, \mathfrak{h}_{\mathbf{R}})^\wedge} [e^{-I_{P_1}(\phi_{P_1}(f, f_1, \ldots, f_N, g))} e^{-I_{P_2}(\phi_{P_2}(g, h_1, \ldots, h_M))} dg]$$

$$= K_{P_1} K_{P_2} \int_{C^\infty(\mathbf{S}^1, \mathfrak{h}_{\mathbf{R}})^\wedge} [e^{-I_{P_3}(\widetilde{\widetilde{g}})} dg] \cdot e^{-I_{P_3}(\phi_{P_3}(f, f_1, \ldots, f_N, h_1, \ldots, h_M))}.$$

On the other hand, we have
$$K_{P_3}(f, f_1, \ldots, f_N, h_1, \ldots, h_M) = K_{P_3} \, e^{-I_{P_3}(\phi_{P_3}(f, f_1, \ldots, f_N, h_1, \ldots, h_M))}.$$
So the theorem is proved.

(2) In the same way, by Proposition (6.8), we have

$$\int_{C^\infty(\mathbf{S}^1, \mathbf{T}_L)^\wedge} K_{P_1, L}(f, f_1, \ldots, f_N, g) \, K_{P_2, L}(g, h_1, \ldots, h_M)[dg]$$

$$= K_{P_1} K_{P_2} \int_{C^\infty(\mathbf{S}^1, \mathbf{T}_L)^\wedge} \sum_t e^{-I_{P_1}(\phi_{P_1}(f, f_1, \ldots, f_N, g)_t)} \sum_s e^{-I_{P_2}(\phi_{P_2}(g, h_1, \ldots, h_M)_s)}[dg]$$

$$= K_{P_1} K_{P_2} \int_{C^\infty(\mathbf{S}^1, \mathfrak{h}_{\mathbf{R}})^\wedge} [e^{-I_{P_3}(\widetilde{\widetilde{g}})} dg] \cdot \sum_u e^{-I_{P_3}(\phi_{P_3}(f, f_1, \ldots, f_N, h_1, \ldots, h_M)_u)}.$$

On the other hand, we have
$$K_{P_3, L}(f, f_1, \ldots, f_N, h_1, \ldots, h_M) = K_{P_3} \sum_u e^{-I_{P_3}(\phi_{P_3}(f, f_1, \ldots, f_N, h_1, \ldots, h_M)_u)}.$$
So the theorem is proved.

THEOREM (6.21). [Geometric Realization of Associative Law of Neutral Vertex Operators]
We have

$$< v' \mid Y(u, z) Y(v, w) \cdot v'' > = < v' \mid Y(Y(u, z - w)v, w)v'' >,$$

for $u, v \, v', v'' \in U_{L'}$. (See Proposition (2.26).)

Proof : Let us consider the following two diagrams, $P_1 = P_{0,q',w,\rho';q}$ sewed with $P_2 = P_{0,q,z,\rho;1}$, and $P_3 = P_{0,\rho',z-w,\rho;q''}$ sewed with $P_4 = P_{0,q',w,q'';1}$.

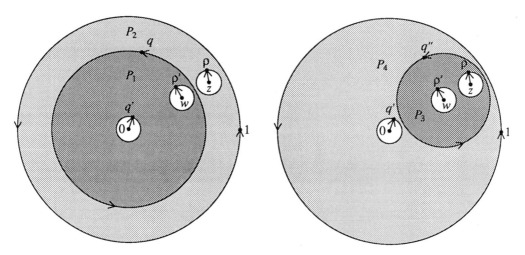

Since the both sewings give the same 3-holed disk $P_{0,q',w,\rho',z,\rho;1}$, it follows from Theorem (6.20) that

$$\int_{C^\infty(\mathbf{S}^1,\mathbf{T}_L)^\wedge} K_{P_2,L}(f;g',g)\, K_{P_1,L}(g;g'',h)[dg]$$

$$= \int_{C^\infty(\mathbf{S}^1,\mathbf{T}_L)^\wedge} K_{P_4,L}(f;f',h)\, K_{P_3,L}(f';g',g'')[df'].$$

For $u, v\ v', v'' \in U_{L'}$, by taking the integral

$$\int_{(C^\infty(\mathbf{S}^1,\mathbf{T}_L)^\wedge)^4} \cdots v'(f)\, e^{-\frac{1}{2}(f_*,\overline{f}_*)} u(g')\, e^{-\frac{1}{2}(g'_*,\overline{g}'_*)}$$

$$\cdot v(g'')\, e^{-\frac{1}{2}(g''_*,\overline{g}''_*)} v''(h)\, e^{-\frac{1}{2}(h_*,\overline{h}_*)} [df\, dg'\, dg''\, dh]$$

of both sides, it follows from Theorem (6.19) that

$$<v'\mid Y(\rho^{L_0}\overline{\rho}^{\overline{L}_0}u,z)q^{L_0}\overline{q}^{\overline{L}_0}\cdot q^{-L_0}\overline{q}^{-\overline{L}_0}Y(\rho'^{L_0}\overline{\rho}'^{\overline{L}_0}v,w)q'^{L_0}\overline{q}'^{\overline{L}_0}\cdot v''>$$

$$=<v'\mid Y(q''^{L_0}\overline{q}''^{\overline{L}_0}\cdot q''^{-L_0}\overline{q}''^{-\overline{L}_0}Y(\rho^{L_0}\overline{\rho}^{\overline{L}_0}u,z-w)\rho'^{L_0}\overline{\rho}'^{\overline{L}_0}v,w)q'^{L_0}\overline{q}'^{\overline{L}_0}\cdot v''>.$$

Namely,

$$<v'\mid Y(\rho^{L_0}\overline{\rho}^{\overline{L}_0}u,z)Y(\rho'^{L_0}\overline{\rho}'^{\overline{L}_0}v,w)q'^{L_0}\overline{q}'^{\overline{L}_0}\cdot v''>$$

$$=<v'\mid Y(Y(\rho^{L_0}\overline{\rho}^{\overline{L}_0}u,z-w)\rho'^{L_0}\overline{\rho}'^{\overline{L}_0}v,w)q'^{L_0}\overline{q}'^{\overline{L}_0}\cdot v''>.$$

Taking the limits of $\rho, \rho', q' \to 1$, we have

$$<v'\mid Y(u,z)Y(v,w)\cdot v''>=<v'\mid Y(Y(u,z-w)v,w)\cdot v''>.$$

Remark. The associativity of neutral vertex operators implies that we have

$$| <v' \mid Y(u,z)Y(v,w) \cdot v'' > | = | <v' \mid Y(Y(u,z-w)v,w) \cdot v'' > |,$$

where $u, v \in V_L$ and $v', v'' \in V_{L'}$. This implies the associativity of vertex operators if we use the fact that $<v' \mid Y(u,z)Y(v,w) \cdot v'' >$ and $<v' \mid Y(Y(u,z-w)v,w) \cdot v'' >$ are meromorphic functions (Proposition (2.16)).

PROPOSITION **(6.22)**. [*N*-holed Disk and Product of Vertex Operators]
Let $P = P_{z_1,\rho_1,\ldots,z_N,\rho_N;q'}$ be an N-holed disk. Then the functional $K_{P,L}(f,g_1,\ldots,g_N)$ is the integral kernel of an operator $\mathcal{K} \otimes \cdots \otimes \mathcal{K} \to \mathcal{K}$ which is the extension of the operator

$$q^{-L_0}\overline{q}^{-\overline{L}_0}Y(\rho_1^{L_0}\overline{\rho}_1^{\overline{L}_0}(\;),z_1)\cdots Y(\rho_N^{L_0}\overline{\rho}_N^{\overline{L}_0}(\;),z_N) : U_{L'} \otimes \cdots \otimes U_{L'} \to \mathcal{K}.$$

Namely we have

$$\int_{C^\infty(\mathbf{S}^1,\mathbf{T}_L)^{\wedge}} \cdots \int_{C^\infty(\mathbf{S}^1,\mathbf{T}_L)^{\wedge}} K_{P,L}(f,g_1,\ldots,g_N)$$

$$\cdot u_1(g_1)\,e^{-\frac{1}{2}(g_{1*},\overline{g}_{1*})}\cdots u_N(g_N)\,e^{-\frac{1}{2}(g_{N*},\overline{g}_{N*})}[dg_1]\cdots[dg_N]$$

$$= \left(q^{-L_0}\overline{q}^{-\overline{L}_0}Y(\rho_1^{L_0}\overline{\rho}_1^{\overline{L}_0}u_1,z_1)\cdots Y(\rho_N^{L_0}\overline{\rho}_N^{\overline{L}_0}u_N,z_N)\cdot e^{(0,0)}\right)(f)\,e^{-\frac{1}{2}(f_*,\overline{f}_*)},$$

for all $u_1,\ldots,u_N \in U_{L'}$.

Proof : The statement follows from Theorems (6.17) and (6.19) by sewing $N-1$ of 2-holed disks and one 1-holed disk.

CHAPTER III

ANALYTIC REALIZATION OF VERTEX OPERATOR ALGEBRAS

§7. Zeta-Regularization

§8. Zeta-Regularized Determinants on Cylinders and Elliptic Curves

§9. String Path Integrals over Cylinders and Elliptic Curves

In this chapter, we define and compute string path integrals on cylinders and elliptic curves using the method of zeta-regularization. We show that the action of the vertex operators can be realized by them.

In section 7, we define the notion of zeta-regularized infinite product and prove its several basic properties. In section 8, we compute zeta-regularized determinants of the Laplacians on cylinders and elliptic curves. We show that the trace $\mathrm{Tr}\ q^d \bar{q}^{\bar{d}}$ can be expressed using the zeta-regularized determinant of the Laplacian on an elliptic curve E_τ ($q = e^{2\pi i \tau}$) and thus the sum $\sum_{\omega \in L'/L} |\widetilde{ch}_q V_{(\omega)}|^2$ is modular invariant. In section 9, we define and compute string path integrals over functions on elliptic curves. We show that the trace $\mathrm{Tr}\ \widetilde{Y}(e^{(\alpha_1, \alpha_1)}, z_1) \cdots \widetilde{Y}(e^{(\alpha_N, \alpha_N)}, z_N) \cdot q^d \bar{q}^{\bar{d}}$ can be realized as string path integrals. Also, we define and compute string path integrals over functions on cylinders and we show that they provide integral kernel of products of the neutral vertex operators. Thus we obtain an analytic realization of the actions of the vertex operators. When L is a root lattice of ADE-type, we construct all the level $= 1$ standard representations of affine Lie algebra of $\widehat{A}\widehat{D}\widehat{E}$-type analytically.

§7. Zeta-Regularization.

§7-A. Zeta-Regularized Infinite Products.

We define a zeta-regularized infinite product and examine its basic properties.

DEFINITION **(7.1)**. [Zeta Regularization]
For a sequence $\Lambda = \{\lambda_1, \lambda_2, \ldots\}$ of positive real numbers, we define a function of s by

$$\zeta_\Lambda(s) = \sum_{n=1}^{\infty} \frac{1}{\lambda_n{}^s}.$$

We assume that $\zeta_\Lambda(s)$ converges when the real part of s is sufficiently large. If $\zeta_\Lambda(s)$ can be analytically continued and is regular at $s = 0$, we define the zeta-regularized product by

$$\left(\prod_{n=1}^{\infty} \lambda_n \right)_\zeta = \exp(-\zeta_\Lambda{}'(0)).$$

We also say that the sequence Λ is zeta-multipliable.

This definition is, of course, motivated by the fact that

$$\prod_{n=1}^{N} \lambda_n = \exp\left(-\frac{d}{ds}\bigg|_{s=0} \sum_{n=1}^{N} \frac{1}{\lambda_n{}^s} \right),$$

when the sequence is finite.

Remark. If the usual limit $\prod_{n=1}^{\infty} \lambda_n$ exists, then $\lim_{n \to \infty} \lambda_n = 1$, thus the zeta function does not converge anywhere.

Example **(7.2)**. [Riemann Zeta Function]
For the sequence $\{1, 2, 3, \ldots\}$, the corresponding zeta function is the Riemann zeta function $\zeta(s)$.

$$\zeta(s) = \sum_{n=1}^{\infty} \frac{1}{n^s},$$

for $\text{Re}\, s > 1$. ζ can be analytically continued to a meromorphic function on the complex plane $\mathbb{C}$.

(1) $\zeta(s)$ has a simple pole at $s = 1$ and

$$\zeta(s) = \frac{1}{s-1} + \gamma + a_1(s-1) + a_2(s-1)^2 + \cdots$$

for $s \approx 1$, where γ is the Euler constant.

(2) $\zeta(0) = -\frac{1}{2}$.

(3) $\dfrac{\Gamma(s/2)}{\pi^{s/2}}\zeta(s) = \dfrac{\Gamma((1-s)/2)}{\pi^{(1-s)/2}}\zeta(1-s)$.

(4) $\zeta'(0) = -\log\sqrt{2\pi}$. Therefore we have

$$\left(\prod_{n=1}^{\infty} n\right)_{\zeta} = \sqrt{2\pi}.$$

Example (7.3). [Epstein Zeta Function, Kronecker's First Limit Formula] (See Kronecker [Kr], Siegel [Si].)

Let $\tau = \tau_1 + i\tau_2$ be a complex number such that $\tau_2 > 0$. For the sequence $\left\{\frac{|n+m\tau|^2}{\tau_2}\right\}_{\substack{n,m\in\mathbb{Z} \\ (n,m)\neq(0,0)}}$, the corresponding zeta function is the Epstein zeta function

$$\zeta_\tau(s) = \sum_{\substack{n,m\in\mathbb{Z} \\ (n,m)\neq(0,0)}} \left(\frac{\tau_2}{|n+m\tau|^2}\right)^s$$

for $\mathrm{Re}\,s > 1$. ζ_τ can be analytically continued to a meromorphic function on the complex plane $\mathbb{C}$.

(1) $\zeta_\tau(s)$ has a simple pole only at $s = 1$ and

$$\zeta_\tau(s) = \frac{\pi}{s-1} + 2\pi(\gamma - \log(2\sqrt{\tau_2}|\eta(\tau)|^2)) + a_1(s-1) + a_2(s-1)^2 + \cdots$$

for $s \approx 1$, where γ is the Euler constant.

(2) $\zeta_\tau(0) = -1$.

(3) $\dfrac{\Gamma(s)}{\pi^s}\zeta_\tau(s) = \dfrac{\Gamma(1-s)}{\pi^{1-s}}\zeta_\tau(1-s)$.

(4) $\zeta_\tau'(0) = -\log(4\pi^2\tau_2|\eta(\tau)|^4)$. Therefore we have

$$\left(\prod_{\substack{n,m\in\mathbb{Z} \\ (n,m)\neq(0,0)}} \frac{|n+m\tau|^2}{\tau_2}\right)_{\zeta} = 4\pi^2\tau_2|\eta(\tau)|^4.$$

where $\eta(\tau)$ is the eta function.

PROPOSITION (7.4).

Let $\Lambda = \{\lambda_1, \lambda_2, \ldots\}$ be a zeta-multipliable sequence of positive real numbers.

(1) For a positive real number α, we have

$$\left(\prod_{n=1}^{\infty} \lambda_n{}^{\alpha}\right)_{\varsigma} = \left\{\left(\prod_{n=1}^{\infty} \lambda_n\right)_{\varsigma}\right\}^{\alpha}.$$

(2) For a positive real number μ, we have

$$\left(\prod_{n=1}^{\infty} \lambda_n \mu\right)_{\varsigma} = \left(\prod_{n=1}^{\infty} \lambda_n\right)_{\varsigma} \mu^{\zeta_\Lambda(0)}.$$

Remark : Note that in the second equation, the value $\zeta_\Lambda(0)$ plays the role of generalized dimension, since in the finite-dimensional case we have

$$\prod_{n=1}^{N} \lambda_n \mu = \left(\prod_{n=1}^{N} \lambda_n\right) \mu^{N}.$$

PROPOSITION (7.5).

Let $\Lambda = \{\lambda_1, \lambda_2, \ldots\}$ be a zeta-multipliable sequence of positive real numbers and let $A = \{\alpha_1, \alpha_2, \ldots\}$ be a sequence of positive real numbers such that the limit $\sum_{n=1}^{\infty} |\log \alpha_n|$ exists. Then the sequence

$$\Lambda \cdot A = \{\lambda_1 \alpha_1, \lambda_2 \alpha_2, \ldots\}$$

is zeta-multipliable and we have

$$\zeta_{\Lambda \cdot A}(0) = \zeta_\Lambda(0),$$

$$\zeta_{\Lambda \cdot A}{}'(0) = \zeta_\Lambda{}'(0) - \sum_{n=1}^{\infty} \log \alpha_n,$$

$$\left(\prod_{n=1}^{\infty} \lambda_n \alpha_n\right)_{\varsigma} = \left(\prod_{n=1}^{\infty} \lambda_n\right)_{\varsigma} \prod_{n=1}^{\infty} \alpha_n.$$

§7-B. Zeta-Regularized Determinant of Laplacian and Conformal Anomaly.

DEFINITION (7.6).

Let X be a Riemann surface, and let E be a vector bundle over X. The smooth sections of E over X is denoted by $C^\infty(X, E)$. We define a map

$$T : C^\infty(X, E) \to C^\infty(\partial X, E^2)$$

by

$$T(\phi) = \left(\phi\big|_{\partial X}, \frac{\partial \phi}{\partial v}\Big|_{\partial X} \right).$$

Let $L : C^\infty(X, E) \to C^\infty(X, E)$ be a positive elliptic partial differential operator. A boundary condition is a subspace B of $C^\infty(\partial X, E^2)$. The operator L restricted to the space $T^{-1}(B)$ is denoted by L_B.

THEOREM (7.7). (Seeley [See2])

Under the assumption of Definition (7.6), let Λ_L be the positive eigenvalues of L, then Λ_L is zeta multipliable sequence. The zeta function of Λ_L is called the zeta function of the operator L. We define the zeta-regularized determinant of the operator L by

$$\det{}_\zeta(L) = \exp(-\zeta'_{\Lambda_L}(0)).$$

DEFINITION (7.8). [Zeta Function of Riemann Surface]

Let X be a Riemann surface with a compatible metric g.

(1) If X is closed, we define ζ_X as the zeta function of the operator $-\Delta_g$ over the space $C^\infty{}_*(X, \mathbb{R})$.

(2) If X is open, we define ζ_X as the zeta function of the operator $-\Delta_g$, where Δ_g is the Laplacian with the Dirichlet boundary condition. Namely $-\Delta_g$ over the space $C^\infty(0, \ldots, 0) \subset C^\infty(X, \mathbb{R})$.

THEOREM (7.9).

(1) If X is closed, then we have

$$\zeta_X(0) = \frac{\chi(X)}{6} - 1,$$

where $\chi(X)$ is the Euler characteristics of X. The value $\zeta_X(0)$ does not depend on the choice of the metric g. In fact, it is a topological invariant.

(2) If X is open, then we have

$$\zeta_X(0) = \frac{1}{12\pi} \iint_X K\omega - \frac{1}{24\pi} \int_{\partial X} Js,$$

where K is the Gaussian curvature of X, and ω is the area element, J is the geodesic curvature of the boundary ∂X in X, and s is the length element of ∂X. In general, $\zeta_X(0)$ depends on the choice of the metric g.

Proof : Define

$$Y_X(t) = \mathrm{Tr}(e^{t\Delta}), \quad t > 0,$$

where the trace is over the space $C^\infty_*(X, \mathbb{R})$, when X is closed, and over the space $C^\infty_X(0,\ldots,0)$, when X is open. Then we have,

$$\zeta_X(s) = \frac{1}{\Gamma(s)} \int_0^\infty t^{s-1} Y_X(t) dt,$$

for $\mathrm{Re}(s)$ sufficiently large.

If we have the asymptotic expansion

$$Y_X(t) = \sum_{k>0} a_{-k} t_{-k} + a_0 + \cdots, \qquad (t \to 0^+),$$

where the sum is over a finite number of terms, then we have

$$\zeta_X(0) = a_0.$$

(See [SchwarzA].) The expansion is computed in McKean-Singer [MS], and we have

$$Y_X(t) = \frac{1}{4\pi t} Area_g(X) + \left(\frac{\chi(X)}{6} - 1\right) + \cdots,$$

when X is closed and

$$Y_X(t) = \frac{1}{4\pi t} Area_g(X) - \frac{1}{8\sqrt{\pi t}} Length(\partial X) + \left(\frac{1}{12\pi} \iint_X K\omega - \frac{1}{24\pi} \int_{\partial X} Js\right) + \cdots,$$

when X is open. Thus we get the theorem.

Example **(7.10).**

(1) For the sphere $\mathbf{S}^2$, we have $\quad \chi(\mathbf{S}^2) = 2, \zeta_{\mathbf{S}^2}(0) = -\frac{2}{3}$.

(2) For the disk D, we have $\quad K = 0, \int_{\partial D} Js = 2\pi, \zeta_D(0) = -\frac{1}{12}$.

(3) For the cylinder C, we have $\quad K = J = 0, \zeta_C(0) = 0$.

(4) For the elliptic curve E, we have $\quad \chi(E) = 0, \zeta_E(0) = -1$.

THEOREM **(7.11).** [Conformal Anomaly]

Let g_0 and g_1 be two compatible metrics of a Riemann surface X such that

$$g_1 = e^{2\phi} g_0,$$

where ϕ is not a constant function. Let us put

$$\zeta_i = \text{the zeta function of } X \text{ with metric } g_i. \quad (i = 0, 1.)$$

(1) If X is closed, then we have

$$\zeta_1'(0) - \zeta_0'(0) = \frac{1}{6\pi} \iint_X \phi K_0 \omega_0 + \frac{1}{12\pi} \ll \phi, \phi \gg.$$

(2) If X is open, then we have

$$\zeta_1'(0) - \zeta_0'(0) = \frac{1}{6\pi} \iint\limits_X \phi K_0 \omega_0 - \frac{1}{12\pi} \int\limits_{\partial X} \phi J_0 s_0 + \frac{1}{12\pi} \ll \phi, \phi \gg.$$

Remark : These formulas are known to physicists as the conformal anomaly formulas. (See [Poly].)

Proof : Let us introduce a parameter $u \in [0,1]$ and parametrize metrics by

$$g_u = e^{2u\phi} g_0.$$

Then we have,

$$\Delta_u = e^{-2u\phi} \Delta_0,$$
$$\omega_u = e^{2u\phi} \omega_0,$$
$$K_u = e^{-2u\phi} K_0 - e^{-2u\phi} u \Delta_0 \phi,$$
$$J_u s_u = -2u * d\phi + J_0 s_0.$$

We have

$$\frac{\partial}{\partial u} \zeta_u(s) = \frac{1}{\Gamma(s)} \int\limits_0^\infty t^s \operatorname{Tr}\left(\frac{d\Delta_u}{du} \cdot e^{t\Delta_u} \right) dt = \frac{1}{\Gamma(s)} \int\limits_0^\infty t^s \left(\frac{d}{dt} \operatorname{Tr}(-2\phi e^{t\Delta_u}) \right) dt,$$

for $\operatorname{Re}(s)$ sufficiently large. If we have the asymptotic expansion

$$\operatorname{Tr}(-2\phi e^{t\Delta_u}) = \sum_{k>0} b_{-k} t^{-k} + b_0 + \cdots, \qquad (t \to 0^+),$$

where the sum is over a finite number of terms, then we have

$$\frac{\partial}{\partial u} \zeta_u(0) = 0 \quad \text{and} \quad \frac{\partial}{\partial u} \zeta_u'(0) = -b_0.$$

(See also [SchwarzA].)

(1) If X is closed, then we have

$$b_0 = -\frac{1}{6\pi} \iint\limits_X \phi K_u \omega_u.$$

Using $K_u \omega_u = K_0 \omega_0 - u(\Delta_0 \phi)\omega_0$, we get

$$\frac{\partial}{\partial u} \zeta_u'(0) = \frac{1}{6\pi} \iint\limits_X \phi K_0 \omega_0 + \frac{1}{6\pi} \ll \phi, \phi \gg u.$$

(2) If X is open, then we have

$$b_0 = -\frac{1}{6\pi} \iint\limits_X \phi K_u \omega_u + \frac{1}{12\pi} \int\limits_{\partial X} \phi J_u s_u.$$

Using $J_u s_u = -2u * d\phi + J_0 s_0$, we get

$$\frac{\partial}{\partial u} \zeta_u'(0) = \frac{1}{6\pi} \iint\limits_X \phi K_0 \omega_0 - \frac{1}{12\pi} \int\limits_{\partial X} \phi J_0 s_0 + \frac{1}{6\pi} \ll \phi, \phi \gg u.$$

By integrating these formulas, we get the theorem. (See McKean-Singer [MS], also Bost [Bos].)

§8. Zeta-Regularized Determinants on Cylinders and Elliptic Curves.

In this section, we define string path integrals over cylinders and elliptic curves using the zeta-regularized determinants of Laplacians.

§8-A. Zeta-Regularized Determinants on Elliptic Curves.

We compute the zeta-regularized determinants of the Laplacians on elliptic curves using the Kronecker's first limit formula.

DEFINITION (8.1).

Let L be an even lattice of rank ℓ with a positive biadditive form $< , >$ as in chapter I and II. We put $\mathfrak{h}_{\mathbb{R}} = L \otimes_{\mathbb{Z}} \mathbb{R}$ and $\mathfrak{h} = L \otimes_{\mathbb{Z}} \mathbb{C}$. We extend $< , >$ to $\mathfrak{h}_{\mathbb{R}}$ and $\mathfrak{h}$. We define a new lattice

$$\Gamma = \frac{1}{\sqrt{2}} L.$$

More precisely, $\Gamma = L$ as groups, and the biadditive form $< , >_\Gamma$ of Γ is given by

$$<\alpha, \beta>_\Gamma = \frac{1}{2} <\alpha, \beta>.$$

We extend $< >_\Gamma$ to $\mathfrak{h}_{\mathbb{R}}$ and $\mathfrak{h}$, too. We define a torus

$$\mathbf{T}_L = \mathfrak{h}_{\mathbb{R}}/2\pi L.$$

Note that as a manifold, $\mathbf{T}_L$ is identical to $\mathfrak{h}_{\mathbb{R}}/2\pi\Gamma$.

Let $\tau = \tau_1 + i\tau_2$ be a complex number such that $\tau_2 > 0$ and let E_τ be the elliptic curve associated with τ. (See Example (5.23).) Namely

$$E_\tau = \mathbb{C}/(2\pi\mathbb{Z} + 2\pi\tau\mathbb{Z}).$$

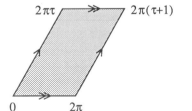
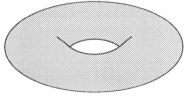

(1) We define a constant Z_{E_τ} by

$$Z_{E_\tau} = \frac{1}{\sqrt{\det_\zeta(-A\Delta)}^\ell},$$

where $A = Area(E_\tau)$ and Δ is the Laplacian on $C^\infty_*(E_\tau, \mathbb{R})$. We regard Z_{E_τ} as a string path integral

$$\int_{C^\infty_*(E_\tau, \mathfrak{h}_\mathbb{R})} e^{-I(\phi)}[d\phi].$$

This is because formally we have

$$\int_{C^\infty_*(E_\tau, \mathfrak{h}_\mathbb{R})} e^{-I(\phi)}[d\phi] = \int_{C^\infty_*(\mathbf{S}^1, \mathfrak{h}_\mathbb{R})} e^{-\frac{1}{4\pi}\iint <\phi,(-\Delta)\phi> r\, dx\, dy}[d\phi] = \frac{1}{\sqrt{\det(-A\Delta)}^\ell}.$$

(2) Recall that any map $\phi \in C^\infty(E_\tau, \mathbf{T}_L)$ can be expressed as

$$\phi = \phi_* + \phi_{0,0} + S_{\alpha,\beta},$$

where $\phi_* \in C^\infty_*(\mathbf{S}^1, \mathfrak{h}_\mathbb{R})$, $\phi_{0,0} \in \mathbf{T}_L$ and

$$S_{\alpha,\beta}(z) = \alpha\left(x - \frac{\tau_1}{\tau_2}y\right) + \beta\frac{y}{\tau_2},$$

where $z = e^{i(x+iy)}$, $\alpha, \beta \in L$. (See Example (5.23).) We have

$$I(\phi) = I(\phi_*) + I(S_{\alpha,\beta}).$$

We define a constant $Z_{E_\tau, L}$ by

$$Z_{E_\tau, L} = Z_{E_\tau} \sum_{\phi_0} e^{-I_{E_\tau}(\phi_0)} \cdot \mathrm{vol}(\mathfrak{h}_\mathbb{R}/2\pi\Gamma).$$

where the sum is over all harmonic maps of form $S_{\alpha,\beta}$. We regard $Z_{E_\tau, L}$ as a string path integral

$$\int_{C^\infty(E_\tau, \mathbf{T}_L)} e^{-I(\phi)}[d\phi].$$

(3) We define a constant $\widetilde{Z}_{E_\tau, L}$ by

$$\widetilde{Z}_{E_\tau, L} = Z_{E_\tau} \sum_{\phi_0} (-1)^{\phi_0} e^{-I(\phi_0)} \cdot \mathrm{vol}(\mathfrak{h}_\mathbb{R}/2\pi\Gamma).$$

where the sum is over all harmonic maps of form $S_{\alpha,\beta}$, and

$$(-1)^{\phi_0} = (-1)^{<\alpha,\beta>} \in \{\pm 1\}.$$

Note that

$$(-1)^{<\alpha,\beta>} = \varepsilon(\alpha,\beta)\varepsilon(\beta,\alpha).$$

We regard $\widetilde{Z}_{E_\tau,L}$ as a string path integral

$$\int_{C^\infty(E_\tau,\mathbf{T}_L)} (-1)^\phi e^{-I(\phi)}[d\phi].$$

Remark. Note that these constants $Z_{E_\tau}, Z_{E_\tau,L}$, and $\widetilde{Z}_{E_\tau,L}$ are modular invariant in τ, because E_τ and the zeta-regularized determinant, $I(\phi_0)$, $(-1)^{\phi_0}$ are all modular invariant notions.

THEOREM **(8.2)**. [Computations of $Z_{E_\tau}, Z_{E_\tau,L}$, and $\widetilde{Z}_{E_\tau,L}$]
We can compute the constants and we have

(1) $Z_{E_\tau} = \dfrac{1}{(\sqrt{\tau_2}|\eta(\tau)|^2)^\ell}.$

(2) $Z_{E_\tau,L} = \dfrac{\displaystyle\sum_{(r,s)\in\Lambda_L} q^{\frac{1}{2}r^2}\overline{q}^{\frac{1}{2}s^2}}{|\eta(\tau)|^{2\ell}}.$

(3) $\widetilde{Z}_{E_\tau,L} = \dfrac{\displaystyle\sum_{(r,s)\in\Omega_L} q^{\frac{1}{2}r^2}\overline{q}^{\frac{1}{2}s^2}}{|\eta(\tau)|^{2\ell}}.$

Λ_L and Ω_L are the lattices defined in Definition (4.22).

Proof : The operator

$$-A\Delta \quad \text{on } C^\infty_*(E_\tau,\mathbb{R})$$

has eigenfunctions

$$\omega_{n,m}(z) = \begin{cases} \sqrt{2}\sin(n(x-\frac{\tau_1}{\tau_2}y)+m\frac{y}{\tau_2}), & n > 0 \text{ or } n=0, m>0, \\ \sqrt{2}\cos(n(x-\frac{\tau_1}{\tau_2}y)+m\frac{y}{\tau_2}), & n < 0 \text{ or } n=0, m<0, \end{cases}$$

$(z = e^{i(x+iy)})$ with eigenvalues $\Lambda_{E_\tau} = \{\lambda_{n,m}\}$,

$$\lambda_{n,m} = \frac{4\pi^2|m-n\tau|^2}{\tau_2}.$$

So, we have

$$\zeta_{\Lambda_{E_\tau}}(s) = \sum_{\substack{n,m\in\mathbb{Z} \\ (n,m)\neq(0,0)}} \frac{1}{\lambda_{n,m}{}^s} = \zeta_\tau(s)(4\pi^2)^{-s}.$$

(Note that $\zeta_{\Lambda_{E_\tau}}(s) = \zeta_E(s)(4\pi^2\tau_2)^{-s}$.) By the Kronecker's first limit formula (Example (7.0).), we have

$$\zeta_{\Lambda_{E_\tau}}{}'(0) = \zeta_\tau{}'(0) + \zeta_\tau(0)(-\log 4\pi^2) = -\log(\tau_2|\eta(\tau)|^4).$$

(1) Therefore we get

$$Z_{E_\tau} = \frac{1}{\sqrt{\det_\zeta(-Area(E_\tau)\Delta)}^\ell} = \frac{1}{(\sqrt{\tau_2}|\eta(\tau)|^2)^\ell}.$$

(2) Using the computations on the harmonic part done in Proposition (A.3), we get

$$Z_{E_\tau,L} = Z_{E_\tau} \sum_{(r,s)\in\Lambda_L} q^{\frac{1}{2}r^2}\overline{q}^{\frac{1}{2}s^2} \cdot \sqrt{\tau_2}^\ell = \frac{\sum\limits_{(r,s)\in\Lambda_L} q^{\frac{1}{2}r^2}\overline{q}^{\frac{1}{2}s^2}}{|\eta(\tau)|^{2\ell}}.$$

(3) In the same way, we get

$$\widetilde{Z}_{E_\tau,L} = Z_{E_\tau} \sum_{(r,s)\in\Omega_L} q^{\frac{1}{2}r^2}\overline{q}^{\frac{1}{2}s^2} \cdot \sqrt{\tau_2}^\ell = \frac{\sum\limits_{(r,s)\in\Omega_L} q^{\frac{1}{2}r^2}\overline{q}^{\frac{1}{2}s^2}}{|\eta(\tau)|^{2\ell}}.$$

PROPOSITION (8.3).

We define the shifted degree operators d and $\overline{d}$ on W (See Proposition (4.23)) by

$$d = L_0 - \frac{\ell}{24},$$
$$\overline{d} = \overline{L}_0 - \frac{\ell}{24}.$$

Then we have

$$\widetilde{Z}_{E_\tau,L} = \text{Tr}_W \, q^d\overline{q}^{\overline{d}} = \sum_{\omega\in L'/L} |\widetilde{ch}_q V_{(\omega)}|^2.$$

COROLLARY (8.4).

$$\sum_{\omega\in L'/L} |\widetilde{ch}_q V_{(\omega)}|^2.$$

is modular invariant function of τ.

Proof : Since $\widetilde{Z}_{E_\tau,L}$ is modular invariant, it is obvious that the above sum is modular invariant. This clarifies the mysterious modular behavior of the characters of the affine Lie algebra representations when L is a root lattice of ADE-type. (See Proposition (1.9).)

§8-B. Zeta-Regularized Determinants on Cylinders.

DEFINITION (8.5).

Let $\tau' = \tau_1' + i\tau_2'$ and $\tau'' = \tau_1'' + i\tau_2''$ be two complex numbers such that $\tau_2'' > \tau_2'$. We put $\tau = \tau'' - \tau'$. Let us take the cylinder $C = C_{\tau',\tau''}$. Although we use the standard metric $du d\bar{u}$ of the cylinder C, we found it better to use the coordinate $z = e^{iu}$ as well as the standard coordinate $u = x + iy$. (See Example (5.21).)

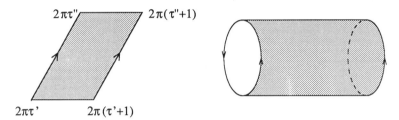

We define

$$K_C = \frac{1}{\sqrt{\det_\zeta(-A\Delta)}^\ell},$$

where $A = Area(C)$ and Δ is the Dirichlet Laplacian on $C^\infty{}_C(0,0) \subset C^\infty(C, \mathbb{R})$. K_C can be regarded as a string path integral

$$\int_{C^\infty{}_C(0,0)} e^{-I(\phi)}[d\phi].$$

This is because formally we have

$$\int_{C^\infty{}_C(0,0)} e^{-I(\phi)}[d\phi] = \int_{C^\infty{}_C(0,0)} e^{-\frac{1}{4\pi}\iint <\phi,(-\Delta)\phi> \mathrm{r}\, dx dy}[d\phi] = \frac{1}{\sqrt{\det(-A\Delta)}^\ell}.$$

(1) Let $f, g \in C^\infty(\mathbf{S}^1, \mathfrak{h}_\mathbb{R})$. We define

$$K_C(f,g) = K_C\, e^{-I(\phi_0)}.$$

where ϕ_0 is the unique harmonic function in $C^\infty{}_C(f,g)$. It can be regarded as a string path integral

$$\int_{C^\infty{}_C(f,g)} e^{-I(\phi)}[d\phi].$$

(2) For $f, g \in C^\infty(\mathbf{S}^1, \mathbf{T}_L)$, we define

$$K_{C,L}(f,g) = K_C \sum_{\phi_0} e^{-I(\phi_0)}.$$

where the sum is over all harmonic maps ϕ_0 in $C^\infty{}_{C,L}(f,g)$. (See Example (5.21).) Note that they are parametrized by $H_1(C,\partial C;L) \cong L$ when $\lambda = \mu$. It can be regarded as a string path integral

$$\int_{C^\infty{}_{C,L}(f,g)} e^{-I(\phi)}[d\phi]$$

(3) For $f,g \in \widetilde{C^\infty}(\mathbf{S}^1,\mathbf{T}_L)$ and for $\phi \in C^\infty{}_{C,L}(f,g)$, define $(-1)^\phi$ in the following way. Express ϕ as

$$\phi = \phi_* + f_0 + S_{\lambda,\beta+(g_o-f_o)/2\pi},$$

where $\phi_* \in C^\infty{}_X(f_*,g_*)$,

$$S_{\lambda,\beta+(g_o-f_o)/2\pi}(z) = \lambda\Big((x - 2\pi\tau_1') - \frac{\tau_1}{\tau_2}(y - 2\pi\tau_2')\Big) + \Big(\beta + \frac{g_0 - f_0}{2\pi}\Big)\frac{y - 2\pi\tau_2'}{\tau_2},$$

and $\beta \in L$. Since $f_0, g_0 \in \mathfrak{h}_{\mathbb{R}}$, β is uniquely determined by ϕ, and we define

$$(-1)^\phi = (-1)^{<\lambda,\beta>}.$$

Note that $(-1)^\phi$ depends only on the harmonic part of ϕ. Using this, we define

$$\widetilde{K}_{C,L}(f,g) = K_C \sum_{\phi_0}(-1)^{\phi_0}e^{-I(\phi_0)},$$

where the sum is again over all harmonic maps ϕ_0 in $C^\infty{}_{C,L}(f,g)$. We regard it as a string path integral

$$\int_{C^\infty{}_{C,L}(f,g)} (-1)^\phi e^{-I(\phi)}[d\phi]$$

PROPOSITION (8.6).
We have

$$K_C = \frac{1}{(\sqrt{2\tau_2}\eta(2i\tau_2))^\ell}.$$

Proof : The operator

$$-A\Delta \quad \text{on } C^\infty{}_C(0,0) \subset C^\infty(C,\mathbb{R})$$

has eigenfunctions

$$\omega_{n,m}(z) = \begin{cases} 2\sin(n(x - 2\pi\tau_1'))\sin(\frac{m(y-2\pi\tau_2')}{2\tau_2}), & \text{for } n > 0, m > 0, \\ \sqrt{2}\sin(\frac{m(y-2\pi\tau_2')}{2\tau_2}), & \text{for } n = 0, m > 0, \\ 2\cos(n(x - 2\pi\tau_1'))\sin(\frac{m(y-2\pi\tau_2')}{2\tau_2}), & \text{for } n < 0, m > 0, \end{cases}$$

with eigenvalues $\Lambda_C = \{\lambda_{n,m}\}$,

$$\lambda_{n,m} = \frac{2\pi^2|m + n \cdot 2i\tau_2|^2}{2\tau_2}.$$

So the zeta function of the operator is

$$\zeta_{\Lambda_C}(s) = \sum_{n \in \mathbb{Z}, m \geq 1} \left(\frac{2\tau_2}{2\pi^2 |m + n \cdot 2i\tau_2|^2} \right)^s.$$

(Note that $\zeta_{\Lambda_C}(s) = \zeta_C(s)(4\pi^2\tau_2)^{-s}$.) Now we can see that

$$\zeta_{\Lambda_C}(s) = \frac{1}{2} \sum_{\substack{n,m \in \mathbb{Z} \\ (n,m) \neq (0,0)}} \left(\frac{2\tau_2}{2\pi^2 |m + n \cdot 2i\tau_2|^2} \right)^s - \sum_{n=1}^{\infty} \left(\frac{2\tau_2}{2\pi^2 |n \cdot 2i\tau_2|^2} \right)^s$$

$$= \frac{1}{2} \zeta_{2i\tau_2}(s)(2\pi^2)^{-s} - \zeta(2s)(4\pi^2\tau_2)^{-s}.$$

Here $\zeta_{2i\tau_2}$ is the Epstein zeta function, and ζ is the Riemann zeta function. Using Examples (7.2) and (7.3), we get

$$\zeta_{\Lambda_C}{}'(0) = \tfrac{1}{2}\zeta_{2i\tau_2}{}'(0) + \tfrac{1}{2}\zeta_{2i\tau_2}(0)(-\log(2\pi^2)) - 2\zeta'(0) - \zeta(0)(-\log(4\pi^2\tau_2))$$

$$= -\tfrac{1}{2}\log(4\pi^2 \cdot 2\tau_2 |\eta(2i\tau_2)|^4) + \tfrac{1}{2}\log(2\pi^2) + 2\log\sqrt{2\pi} - \tfrac{1}{2}\log(4\pi^2\tau_2)$$

$$= -\log(2\tau_2\eta(2i\tau_2)^2).$$

So we have

$$K_C = \frac{1}{\sqrt{\det_\zeta(-A\Delta)}^\ell} = \frac{1}{(\sqrt{2\tau_2}\eta(2i\tau_2))^\ell}.$$

PROPOSITION (8.7).

We have

(1) $\displaystyle K_{C,L}(f,g) = K_C(f_*, g_*) \sum_{(r,s) \in \Lambda_L} \left(q^{\frac{1}{2}r^2} \overline{q}^{\frac{1}{2}s^2} \cdot e^{(r,s)}(f_0 + \lambda\theta) \cdot \overline{e^{(r,s)}}(g_0 + \mu\theta) \right) \frac{\sqrt{2\tau_2}^{-\ell}}{v_L}$,

(2) $\displaystyle \widetilde{K}_{C,L}(f,g) = K_C(f_*, g_*) \sum_{(r,s) \in \Omega_L} \left(q^{\frac{1}{2}r^2} \overline{q}^{\frac{1}{2}s^2} \cdot e^{(r,s)}(f_0 + \lambda\theta) \cdot \overline{e^{(r,s)}}(g_0 + \mu\theta) \right) \frac{\sqrt{2\tau_2}^{-\ell}}{v_L}$,

where $f = f_* + f_0 + \lambda\theta$, $g = g_* + g_0 + \mu\theta$.

Proof: First, note that the harmonic map in $C^\infty_{C,L}(f,g)$ does not exist when $\lambda \neq \mu$. Also when $\lambda = \mu$, the harmonic maps in $C^\infty_{C,L}(f,g)$ are written as

$$\phi_0 = \phi_{0*} + f_0 + S_{\mu, \beta + (g_0 - f_0)/2\pi},$$

where $\phi_{0*} = \phi_C(f_*, g_*), \beta \in L$. Furthermore we have

$$I(\phi_0) = I(\phi_{0*}) + I(S_{\mu, \beta + (g_0 - f_0)/2\pi}).$$

Therefore, we get

(1) $K_{C,L}(f,g) = K_{C,L}(f_*, g_*) \cdot \delta_{\lambda,\mu} \cdot \sum_{\beta \in L} e^{-I(S_{\mu,\beta+(g_0-f_0)/2\pi})}.$

(2) $\widetilde{K}_{C,L}(f,g) = K_{C,L}(f_*, g_*) \cdot \delta_{\lambda,\mu} \cdot \sum_{\beta \in L} (-1)^{<\mu,\beta>} e^{-I(S_{\mu,\beta+(g_0-f_0)/2\pi})}.$

Using the calculation done in Proposition (A.3), we obtain the formulas.

COROLLARY (8.8).
We have the following transformation properties.

(1) $\widetilde{K}_{C,L}(f+2\pi\nu, g) = (-1)^{<\lambda,\nu>} \widetilde{K}_{C,L}(f,g),$

$\widetilde{K}_{C,L}(f, g+2\pi\nu) = (-1)^{<\mu,\nu>} \widetilde{K}_{C,L}(f,g),$

(2) $\widetilde{K}_{C,L}(f+2\pi\nu, f+2\pi\nu) = \widetilde{K}_{C,L}(f,f),$

for all $\nu \in L$ when $f = f_* + f_0 + \lambda\theta$, $g = g_* + g_0 + \mu\theta$.

THEOREM (8.9). [String Path Integral Realization of the Operator $q^d \overline{q}^{\overline{d}}$]
The functional $\widetilde{K}_{C,L}(f,g)$ represents operator $q^d \overline{q}^{\overline{d}}$ on W, namely

$$\int_{C^\infty(S^1, \mathbf{T}_L)^\wedge} \widetilde{K}_{C,L}(f,g) v(g) e^{-\frac{1}{2}(g_*, \overline{g}_*)} [dg] = \left(q^d \overline{q}^{\overline{d}} \cdot v\right)(f) e^{-\frac{1}{2}(f_*, \overline{f}_*)}, \quad \text{for } v \in W.$$

Proof : Let us put $v = v_* \otimes e^{(r_0, s_0)}$ where $v_* \in Sym(\widehat{\mathfrak{h}}^\pm)$ and $(r_0, s_0) \in \Omega_L$. Let us take a generating function

$$v_* = \prod_{n=1}^\infty \sum_{m=0}^\infty \frac{1}{m!} (\alpha_n(-n) + \beta_n(n))^m,$$

where $\alpha_n, \beta_n \in \mathfrak{h}$. Note that

$$v_*(g_*) = \prod_{n=1}^\infty \frac{e^{in<\alpha_n, g_n>} e^{in<\beta_n, \overline{g}_n>}}{e^{-n<\alpha_n, \beta_n>}},$$

where $g_*(\theta) = \sum_{n \neq 0} g_n e^{in\theta}, g_n \in \mathfrak{h}, g_{-n} = \overline{g}_n$. Then we have

$$\int_{C^\infty_*(S^1, \mathfrak{h}_\mathbf{R})^\wedge} K_C(f_*, g_*) v_*(g_*) e^{-\frac{1}{2}(g_*, \overline{g}_*)} [dg_* d\overline{g}_*]$$

$$= K_C \int_{C^\infty_*(S^1, \mathfrak{h}_\mathbf{R})^\wedge} e^{-\frac{1}{2} \sum_{n=1}^\infty \frac{n}{1-q^n\overline{q}^n}((1+q^n\overline{q}^n)(f_n\overline{f}_n + g_n\overline{g}_n) - 2q^n f_n\overline{g}_n - 2\overline{q}^n \overline{f}_n g_n)}$$

$$\cdot \prod_{n=1}^\infty \frac{e^{in<\alpha_n, g_n>} e^{in<\beta_n, \overline{g}_n>}}{e^{-n<\alpha_n, \beta_n>}} \cdot \exp\left(-\frac{1}{2} \sum_{n=1}^\infty n <g_n, \overline{g}_n>\right) [dg_* d\overline{g}_*]$$

$$= K_C$$

$$\cdot \prod_{n=1}^{\infty} \int_{\mathfrak{h}} e^{-\frac{1}{2}\frac{n}{1-q^n\overline{q}^n}((1-q^n\overline{q}^n)f_n\overline{f}_n + 2(g_n - q^n f_n)(\overline{g}_n - \overline{q}^n\overline{f}_n))} \cdot \frac{e^{in<\alpha_n,g_n>}e^{in<\beta_n,\overline{g}_n>}}{e^{-n<\alpha_n,\beta_n>}} dg_n d\overline{g}_n$$

$$= K_C \prod_{n=1}^{\infty} \int_{\mathfrak{h}} e^{-\frac{1}{2}\frac{n}{1-q^n\overline{q}^n}((1-q^n\overline{q}^n)f_n\overline{f}_n + 2g_n\overline{g}_n)} \cdot \frac{e^{in<\alpha_n,g_n+q^n f_n>}e^{in<\beta_n,\overline{g}_n+\overline{q}^n\overline{f}_n>}}{e^{-n<\alpha_n,\beta_n>}} dg_n d\overline{g}_n$$

$$= K_C \prod_{n=1}^{\infty} \left\{ \frac{e^{in<\alpha_n,q^n f_n>}e^{in<\beta_n,\overline{q}^n\overline{f}_n>}}{e^{-n<\alpha_n,\beta_n>}} \cdot e^{-\frac{1}{2}n f_n\overline{f}_n} \right.$$

$$\left. \cdot \int_{\mathfrak{h}} e^{in<\alpha_n,g_n>}e^{in<\beta_n,\overline{g}_n>}e^{-\frac{n}{1-q^n\overline{q}^n}g_n\overline{g}_n} dg_n d\overline{g}_n \right\}$$

$$= K_C \prod_{n=1}^{\infty} \left\{ \frac{e^{in<\alpha_n,q^n f_n>}e^{in<\beta_n,\overline{q}^n\overline{f}_n>}}{e^{-n<\alpha_n,\beta_n>}} \cdot e^{-\frac{1}{2}n f_n\overline{f}_n} \cdot e^{-n(1-q^n\overline{q}^n)<\alpha_n,\beta_n>} \right.$$

$$\left. \cdot \int_{\mathfrak{h}} e^{-\frac{n}{1-q^n\overline{q}^n}g_n\overline{g}_n} dg_n d\overline{g}_n \right\}$$

$$= \frac{1}{\sqrt{2\tau_2}^\ell} \frac{1}{(q^{\frac{1}{24}}\overline{q}^{\frac{1}{24}})^\ell} \cdot \prod_{n=1}^{\infty} \frac{e^{in<q^n\alpha_n,f_n>}e^{in<\overline{q}^n\beta_n,\overline{f}_n>}}{e^{-n<q^n\alpha_n,\overline{q}^n\beta_n>}} \cdot e^{-\frac{1}{2}(f_*,\overline{f}_*)}$$

$$= \frac{1}{\sqrt{2\tau_2}^\ell} (q^d\overline{q}^{\overline{d}} \cdot v)(f_*) e^{-\frac{1}{2}(f_*,\overline{f}_*)}.$$

Therefore we have

$$\int_{C^\infty(\mathbf{S}^1, \mathbf{T}_L)^\wedge} \widetilde{K}_{C,L}(f,g)v(g) e^{-\frac{1}{2}(g_*,\overline{g}_*)} [dg]$$

$$= \int_{C^\infty_*(\mathbf{S}^1, \mathfrak{h}_{\mathbb{R}})^\wedge} K_C(f_*, g_*)v_*(g_*) e^{-\frac{1}{2}(g_*,\overline{g}_*)} [dg_* d\overline{g}_*]$$

$$\cdot \sum_{\mu \in L} \int_{\mathbf{T}_L} \sum_{(r,s) \in \Omega_L} \left(q^{\frac{1}{2}r^2}\overline{q}^{\frac{1}{2}s^2} e^{(r,s)}(f_0 + \lambda\theta) \cdot \overline{e^{(r,s)}}(g_0 + \mu\theta) \right) \frac{\sqrt{2\tau_2}^\ell}{v_L} \cdot e^{(r_0,s_0)}(g_0 + \mu\theta) dg_0$$

$$= \left(q^d\overline{q}^{\overline{d}} \cdot v_* \right)(f_*) \cdot e^{-\frac{1}{2}(f_*,\overline{f}_*)} \cdot q^{\frac{1}{2}r_0^2}\overline{q}^{\frac{1}{2}s_0^2} \cdot e^{(r_0,s_0)}(f_0 + \lambda\theta)$$

$$= \left(q^d\overline{q}^{\overline{d}} \cdot v \right)(f) e^{-\frac{1}{2}(f_*,\overline{f}_*)}.$$

Remark. We decompose the operator

$$q^d\overline{q}^{\overline{d}} = e^{2\pi i\tau_1(d-\overline{d})} e^{-2\pi\tau_2(d+\overline{d})},$$

into

$$P_{\tau_1} = e^{2\pi i\tau_1(d-\overline{d})},$$

and

$$H_{\tau_2} = e^{-2\pi\tau_2(d+\overline{d})}.$$

Here the operator

$$d + \overline{d} = L_0 + \overline{L}_0 - \frac{\ell}{12}$$

is the renormalized Hamiltonian, and the operator

$$d - \overline{d} = L_0 - \overline{L}_0$$

is the momentum operator. The operator P_{τ_1} has a nice realization

$$(P_{\tau_1} \cdot v)(f) = v(f_{\tau_1}),$$

for $v \in W$, where the function f_{τ_1} is defined as

$$f_{\tau_1}(\theta) = f(\theta + 2\pi\tau_1).$$

§8-C. Sewing of Determinants on Cylinders.

We prove the sewing property of the zeta-regularized determinants on cylinders.

PROPOSITION (8.10). [Sewing of Determinants]
Let $\tau'' = \tau_1'' + i\tau_2''$, $\tau''' = \tau_1''' + i\tau_2'''$, and $\tau'''' = \tau_1'''' + i\tau_2''''$ be three complex numbers such that $\tau_2'''' > \tau_2''' > \tau_2''$. Let us put $\tau = \tau''' - \tau''$ and $\tau' = \tau'''' - \tau'''$. Let $C_1 = C_{\tau'',\tau'''}$ and $C_2 = C_{\tau''',\tau''''}$ be two cylinders, and let $C_3 = C_{\tau'',\tau''''}$ be the cylinder obtained by sewing the two cylinders C_1 and C_2.

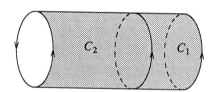

Then we have the following sewing property of zeta-regularized determinants of Laplacians. Namely, we have

$$\frac{1}{\sqrt{\det_\zeta(-A_3\Delta_3)}} = \frac{1}{\sqrt{\det_\zeta(-A_1\Delta_1)}} \frac{1}{\sqrt{\det_\zeta(-A_2\Delta_2)}} \int\limits_{C^\infty(\mathbf{S}^1,\mathbb{R})^\wedge} [e^{-I_{C_3}(\widetilde{\widetilde{g}})} dg],$$

where $A_i = Area(C_i)$ and Δ_i is the Dirichlet Laplacian on C_i,

$$\widetilde{\widetilde{g}} = \begin{cases} \phi_{C_1}(0,g), & \text{on } C_1 \\ \phi_{C_2}(g,0), & \text{on } C_2. \end{cases}$$

In other words, we have

$$K_{C_3} = K_{C_1} K_{C_2} \int\limits_{C^\infty(\mathbf{S}^1,\mathfrak{h}_\mathbb{R})^\wedge} [e^{-I_{C_3}(\widetilde{\widetilde{g}})} dg].$$

Proof: Look at the proof of Proposition (6.4).

PROPOSITION (8.11). [Markov Property]

Under the same assumption of Proposition (8.10), we have the following Markov property or semigroup property.

(1) $\quad \displaystyle\int_{C^\infty(\mathbf{S}^1,\mathfrak{h}_\mathbb{R})^\wedge} K_{C_1}(f,g)\, K_{C_2}(g,h)[dg] = K_{C_3}(f,h), \quad$ for $f,h \in C^\infty(\mathbf{S}^1,\mathfrak{h}_\mathbb{R})$.

(2) $\quad \displaystyle\int_{C^\infty(\mathbf{S}^1,\mathbf{T}_L)^\wedge} K_{C_1,L}(f,g)\, K_{C_2,L}(g,h)[dg] = K_{C_3,L}(f,h), \quad$ for $f,h \in C^\infty(\mathbf{S}^1,\mathbf{T}_L)$.

(3) $\quad \displaystyle\int_{C^\infty(\mathbf{S}^1,\mathbf{T}_L)^\wedge} \widetilde{K}_{C_1,L}(f,g)\, \widetilde{K}_{C_2,L}(g,h)[dg] = \widetilde{K}_{C_3,L}(f,h), \quad$ for $f,h \in \widetilde{C^\infty}(\mathbf{S}^1,\mathbf{T}_L)$.

Proof : We can use our previous results Propositions (6.6) and (6.8) and immediately get the equations (1) and (2). To prove (3), we need to modify Proposition (6.8), but it can be done easily.

PROPOSITION (8.12). [Sewing of Determinants]

Let $\tau' = \tau_1' + i\tau_2'$ and $\tau'' = \tau_1'' + i\tau_2''$ be two complex numbers such that $\tau_2'' > \tau_2'$. We put $\tau = \tau'' - \tau'$. Let us take the cylinder $C = C_{\tau',\tau''}$ (See Example (5.21).) and the elliptic curve $E = E_\tau$. Then we have the following sewing property of zeta-regularized determinants of Laplacians. Namely, we have

$$\frac{1}{\sqrt{\det_\zeta(-A\Delta_E)}} = \frac{1}{\sqrt{\det_\zeta(-A\Delta_C)}} \int_{C^\infty_*(\mathbf{S}^1,\mathbb{R})^\wedge} [e^{-I_C(\widetilde{f}_*)} df_* d\overline{f}_*] \cdot \sqrt{2},$$

In other words, we have

$$Z_{E_\tau} = K_C \int_{C^\infty_*(\mathbf{S}^1,\mathfrak{h}_\mathbb{R})^\wedge} [e^{-I(\widetilde{f}_*)} df_* d\overline{f}_*] \cdot \sqrt{2}^\ell.$$

Proof : From Example (5.21), We know that

$$I(\widetilde{f}_*) = \left(\frac{1-T}{\sqrt{1-T\overline{T}}} f_*, \overline{\frac{1-T}{\sqrt{1-T\overline{T}}} f_*} \right).$$

where $T = T_q$. Therefore we have

$$\int_{C^\infty_*(\mathbf{S}^1,\mathbb{R})^\wedge} [e^{-I(\widetilde{f}_*)} df_* d\overline{f}_*] \quad \left(\det \frac{(1-T)(1-\overline{T})}{1-T\overline{T}} \right)^{-1} = \left(\frac{\eta(2i\tau_2)}{|\eta(\tau)|^2} \right).$$

This implies the equation.

PROPOSITION **(8.13)**.

We have the following.

(1) $\displaystyle\int_{C^{\infty}{}_*(\mathbf{S}^1, \mathfrak{h}_{\mathbb{R}})^{\wedge}} K_C(f_*, f_*)[df_* d\overline{f}_*] \cdot \sqrt{2}^{\ell} = Z_{E_\tau}.$

(2) $\displaystyle\int_{C^{\infty}(\mathbf{S}^1, \mathbf{T}_L)^{\wedge}} K_{C,L}(f, f)[df] = Z_{E_\tau, L}.$

(3) $\displaystyle\int_{C^{\infty}(\mathbf{S}^1, \mathbf{T}_L)^{\wedge}} \widetilde{K}_{C,L}(f, f)[df] = \widetilde{Z}_{E_\tau, L}.$

Proof :

(1) This formula is immediately obtained from Proposition (8.12).

(2) Using the result in Example (5.21), we get

$$\int_{C^{\infty}(\mathbf{S}^1, \mathbf{T}_L)^{\wedge}} K_{C,L}(f, f)[df]$$

$$= K_C \int_{C^{\infty}(\mathbf{S}^1, \mathbf{T}_L)^{\wedge}} \sum_t [e^{-I(\widetilde{f}_t)} df]$$

$$= K_C \int_{C^{\infty}{}_*(\mathbf{S}^1, \mathfrak{h}_{\mathbb{R}})^{\wedge}} [e^{-I(\widetilde{f}_*)} df_* d\overline{f}_*] \cdot \sum_{\alpha, \beta} e^{-I(S_{\alpha, \beta})} \cdot \int_{\mathbf{T}_L} df_0.$$

By the above formula (1) and Theorem (8.2), we can see that the last formula is equal to $Z_{E_\tau, L}$.

(3) The last equation can be proved similarly by taking $(-1)^{\phi}$ into account.

§9. String Path Integrals over Cylinders and Elliptic Curves.

§9-A. String Path Integrals on Elliptic Curve

We define and compute string path integrals using the Kronecker's second limit formula.

DEFINITION (9.1). [String Path Integral on Elliptic Curve]

Let $E = E_\tau$ be an elliptic curve. Let $u_1, \ldots, u_N$ be points on E such that $|z_1| > \cdots > |z_N|$, where $z_j = e^{iu_j}$. Let S_j be the circle given by $\theta \mapsto \theta + u_j$ so that $\theta = 0$ on S_j corresponds to $z = z_j$. Let $\alpha_1, \ldots, \alpha_N \in L'$, such that $\alpha_1 + \cdots \alpha_N = 0$.

(1) We define

$$Z_E(e^{(\alpha_1, \alpha_1)}, z_1) \cdots (e^{(\alpha_N, \alpha_N)}, z_N)$$

$$= \lim_{r \uparrow 1} c \cdot \int_{(C^\infty(S^1, \mathfrak{h}_{\mathbb{R}})^\wedge)^N / \mathfrak{h}_{\mathbb{R}}} \ {}_\bullet^\bullet e^{i<\alpha_1 g_r, f_1>} {}_\bullet^\bullet \cdots {}_\bullet^\bullet e^{i<\alpha_N g_r, f_N>} {}_\bullet^\bullet [e^{-I_E(\Phi(f_1, \ldots, f_N))} df_1 \cdots df_N],$$

where $\Phi = \Phi(f_1, \ldots, f_N)$ is the unique function on E such that $\Phi|_{S_i} = f_i$ which is harmonic on E except on the circles $S_1, \ldots, S_N$. $g_r(\theta) = \sum_{n \in \mathbb{Z}} r^{|n|} e^{in\theta}$ and so g_r approaches the delta function

$$\delta(\theta) = \sum_{n \in \mathbb{Z}} e^{in\theta}.$$

c is a normalization constant,

$$c = Z_E \bigg/ \int_{(C^\infty(S^1, \mathfrak{h}_{\mathbb{R}})^\wedge)^N / \mathfrak{h}_{\mathbb{R}}} [e^{-I_E(\Phi(f_1, \ldots, f_N))} df_1 \cdots df_N].$$

(See Definition (4.6) for the definition of Wick ordering.)

Since

$$\lim_{r \uparrow 1} <\alpha_j g_r, f_j> = <\alpha_j, f_j(0)> = <\alpha_j, \Phi(z_j)>,$$

we regard $Z_E(e^{(\alpha_1, \alpha_1)}, z_1) \cdots (e^{(\alpha_N, \alpha_N)}, z_N)$ as a string path integral symbolically written as

$$\int_{C^\infty_*(E_\tau, \mathfrak{h}_{\mathbb{R}})} \ {}_\bullet^\bullet e^{i<\alpha_1, \phi(z_1)>} {}_\bullet^\bullet \cdots {}_\bullet^\bullet e^{i<\alpha_N, \phi(z_N)>} {}_\bullet^\bullet e^{-I(\phi)} [d\phi].$$

(2) We define

$$Z_{E,L}(e^{(\alpha_1,\alpha_1)},z_1)\cdots(e^{(\alpha_N,\alpha_N)},z_N)$$

$$= Z_E(e^{(\alpha_1,\alpha_1)},z_1)\cdots(e^{(\alpha_N,\alpha_N)},z_N)$$

$$\cdot \sum_{\alpha,\beta\in L}\exp(i<\alpha_1,S_{\alpha,\beta}>)\cdots\exp(i<\alpha_N,S_{\alpha,\beta}>)e^{-I(S_{\alpha,\beta})}\cdot\mathrm{vol}(\mathfrak{h}_\mathbb{R}/2\pi\Gamma).$$

It is regarded as a string path integral

$$\int_{C^\infty(E_\tau,\mathbf{T}_L)} {}_\bullet^\bullet e^{i<\alpha_1,\phi(z_1)>} {}_\bullet^\bullet \cdots {}_\bullet^\bullet e^{i<\alpha_N,\phi(z_N)>} {}_\bullet^\bullet e^{-I(\phi)}[d\phi].$$

(3) We define

$$\widetilde{Z}_{E,L}(e^{(\alpha_1,\alpha_1)},z_1)\cdots(e^{(\alpha_N,\alpha_N)},z_N)$$

$$= Z_E(e^{(\alpha_1,\alpha_1)},z_1)\cdots(e^{(\alpha_N,\alpha_N)},z_N)$$

$$\cdot \sum_{\alpha,\beta\in L}\exp(i<\alpha_1,S_{\alpha,\beta}>)\cdots\exp(i<\alpha_N,S_{\alpha,\beta}>)(-1)^{<\alpha,\beta>}e^{-I(S_{\alpha,\beta})}\cdot\mathrm{vol}(\mathfrak{h}_\mathbb{R}/2\pi\Gamma).$$

It is regarded as a string path integral

$$\int_{C^\infty(E_\tau,\mathbf{T}_L)} {}_\bullet^\bullet e^{i<\alpha_1,\phi(z_1)>} {}_\bullet^\bullet \cdots {}_\bullet^\bullet e^{i<\alpha_N,\phi(z_N)>} {}_\bullet^\bullet (-1)^\phi e^{-I(\phi)}[d\phi].$$

DEFINITION (9.2). [Green Function on Elliptic Curve]
Let w be a point on the elliptic curve E_τ. We define the Green function of E_τ at w by the formula

$$G_w(z) = \lim_{s\to 1}\sum_{\substack{n,m\in\mathbb{Z}\\(n,m)\neq(0,0)}}\frac{1}{\lambda_{n,m}{}^s}\omega_{n,m}(z)\omega_{n,m}(w)$$

$$= \lim_{s\to 1}\sum_{\substack{n,m\in\mathbb{Z}\\(n,m)\neq(0,0)}}\left(\frac{\tau_2}{4\pi^2|m+n\tau|^2}\right)^s e^{i[n((x-p)-\frac{\tau_1}{\tau_2}(y-r))-m\frac{y-r}{\tau_2}]},$$

where $u=x+iy, v=p+ir, z=e^{iu}, w=e^{iv}$, and the limit means taking the analytic continuation in s and setting $s=1$. (See Theorem (8.2) for the definition of $\lambda_{n,m}$ and $\omega_{n,m}$.)

Note that ΔG_w is a constant function. Because $G_w(z)=G_z(w)$, we denote $G_w(z)$ also by $G(z,w)$. We have

$$\ll\phi,G_w\gg=\phi(w).$$

for all $\phi\in C^\infty{}_*(E,\mathbb{R})$. Also we have

$$\ll G_z,G_w\gg=G(z,w),\quad\text{for }z\neq w.$$

LEMMA (9.3). [Kronecker's Second Limit Formula] (See Kronecker [Kr], Siegel [Si].)

Let $\tau = \tau_1 + i\tau_2$ be a complex number such that $\tau_2 > 0$. Let $u, v \in \mathbb{R}$. Define a function

$$\zeta_\tau(s; u, v) = \sum_{\substack{n,m \in \mathbb{Z} \\ (n,m) \neq (0,0)}} \left(\frac{\tau_2}{|m + n\tau|^2} \right)^s e^{2\pi i(nu+mv)}$$

for $\mathrm{Re}\, s > 1$. If $u, v \in \mathbb{Z}$, then we have

$$\zeta_\tau(s; u, v) = \zeta_\tau(s),$$

so without loss of generality, we can assume that $u \notin \mathbb{Z}$ or $v \notin \mathbb{Z}$. Then for fixed u and v, the function $\zeta_\tau(s; u, v)$ can be analytically continued to a holomorphic function of s around $s = 1$. We have

$$\zeta_\tau(1; u, v) = 2\pi^2 \tau_2 v^2 - 2\pi \log \left| \frac{\theta_{11}(u - v\tau, \tau)}{\eta(\tau)} \right|.$$

Here, the theta function θ_{11} is defined by

$$\theta_{11}(x, \tau) = \sum_{n \in \mathbb{Z} + \frac{1}{2}} e^{\pi i n^2 \tau + 2\pi i n(x + \frac{1}{2})}$$

$$= i q^{1/8} \left(\sqrt{z} - \frac{1}{\sqrt{z}} \right) \prod_{n=1}^{\infty} (1 - q^n)(1 - q^n z)(1 - q^n z^{-1}),$$

where $z = e^{2\pi i x}$ and $q = e^{2\pi i \tau}$.

Proof : We can see that when $0 < v < 1$, and $\mathrm{Re}\, s > \frac{1}{2}$, we have

$$\zeta_\tau(s; u, v) = \sum_{m \neq 0} \left(\frac{\tau_2}{|m|^2} \right)^s e^{2\pi i m v} + \sum_{n \neq 0} e^{2\pi i n u} \left\{ \sum_{m=-\infty}^{+\infty} \left(\frac{\tau_2}{|m + n\tau|^2} \right)^s e^{2\pi i m v} \right\}.$$

(See Siegel [Si].) Therefore,

$$\zeta_\tau(1; u, v)$$

$$= \sum_{m \neq 0} \frac{\tau_2}{m^2} e^{2\pi i m v} + \sum_{n \neq 0} e^{2\pi i n u} \left\{ \sum_{m=-\infty}^{+\infty} \frac{\tau_2}{|m + n\tau|^2} e^{2\pi i m v} \right\}$$

$$= \tau_2 \left\{ \frac{1}{2} (2\pi v - \pi)^2 - \frac{\pi^2}{6} \right\}$$

$$+ \sum_{n \neq 0} \frac{1}{n} e^{2\pi i n u} \left\{ \frac{1}{2i} \sum_{m=-\infty}^{+\infty} \left(\frac{1}{m + n\bar{\tau}} - \frac{1}{m + n\tau} \right) e^{2\pi i m v} \right\}$$

$$= 2\pi^2 \tau_2 \left(v^2 + v + \frac{1}{6} \right) + \sum_{n \neq 0} \frac{\pi}{n} e^{2\pi i n u} \left\{ \frac{q^n}{1 - q^n} e^{-2\pi i n \tau v} + \frac{1}{1 - \bar{q}^n} e^{-2\pi i n \bar{\tau} v} \right\}$$

$$= 2\pi^2 \tau_2 \left(v^2 + v + \frac{1}{6} \right) - \pi \sum_{n=1}^{\infty} \left\{ \log(1 - q^n e^{2\pi i(u - \tau v)}) + \log(1 - \bar{q}^n e^{-2\pi i(u - \bar{\tau} v)}) \right\}$$

$$- \pi \sum_{n=0}^{\infty} \left\{ \log(1 - q^n e^{-2\pi i(u - \tau v)}) + \log(1 - \bar{q}^n e^{2\pi i(u - \bar{\tau} v)}) \right\}$$

$$= 2\pi^2 \tau_2 v^2 - 2\pi \log \left| \frac{\theta_{11}(u - v\tau, \tau)}{\eta(\tau)} \right|.$$

PROPOSITION (9.4).

For $z, w \in E_\tau$, $z \neq w$, we have

$$\ll G_z, G_w \gg = \frac{1}{2\pi} \left(\frac{(y - r)^2}{4\pi\tau_2} - \log \left| \frac{\theta_{11}((u - v)/2\pi, \tau)}{\eta(\tau)} \right| \right).$$

where $u = x + iy$, $v = p + ir$, $z = e^{iu}$, $w = e^{iv}$.

Proof : Notice that

$$\ll G_z, G_w \gg = \frac{1}{4\pi^2} \zeta_\tau \left(1; \frac{1}{2\pi} \left((x - p) - \frac{\tau_1}{\tau_2}(y - r) \right), -\frac{1}{2\pi} \frac{y - r}{\tau_2} \right).$$

This immediately implies the equation.

DEFINITION (9.5).

We define the non-singular part of the Green function $\ll G_z, G_w \gg$ by the equality

$$\, \vdots \ll G_z, G_w \gg \vdots \, = \ll G_z, G_w \gg + \frac{1}{2\pi} \log \left| \frac{z - w}{\sqrt{zw}} \right|.$$

Then we can take a limit of $w \to z$, and we define

$$\, \vdots \ll G_z, G_z \gg \vdots \, = \lim_{w \to z} \, \vdots \ll G_z, G_w \gg \vdots \, = -\frac{1}{2\pi} \log \left| \eta(\tau)^2 \right|.$$

THEOREM (9.6). [Computations of String Path Integrals]

Let $u_1, \ldots, u_N$ be points on E_τ such that $|z_1| > \cdots > |z_N|$ where $z_j = e^{iu_j}$. Let $\alpha_1, \ldots, \alpha_N \in L'$ such that $\alpha_1 + \cdots + \alpha_N = 0$.

(1) We have

$$Z_E(e^{(\alpha_1, \alpha_1)}, z_1) \cdots (e^{(\alpha_N, \alpha_N)}, z_N)$$

$$= Z_{E_\tau} \prod_i \exp(-2\pi \, \vdots \ll \alpha_i G_{z_i}, \alpha_i G_{z_i} \gg \vdots) \cdot \prod_{i \neq j} \exp(-2\pi \ll \alpha_i G_{z_i}, \alpha_j G_{z_j} \gg)$$

$$= Z_{E_\tau} \prod_{i < j} \left(\exp \left(-\frac{(y_i - y_j)^2}{4\pi\tau_2} \right) \cdot \left| \chi_\tau(z_i, z_j) \right| \right)^{2 < \alpha_i, \alpha_j >},$$

where $u_j = x_j + iy_j$. Note that

$$\chi_\tau(z, w) = \frac{\theta_{11}((u - v)/2\pi, \tau)}{\eta(\tau)^3} = \frac{z - w}{\sqrt{zw}} \prod_{n=1}^{\infty} \frac{(1 - q^n \frac{z}{w})(1 - q^n \frac{w}{z})}{(1 - q^n)^2},$$

where $q = e^{2\pi i\tau}$, $z = e^{iu}$, $w = e^{iv}$. (See Proposition (2.20).)

(2) We have

$$Z_{E,L}(e^{(\alpha_1,\alpha_1)}, z_1) \cdots (e^{(\alpha_N,\alpha_N)}, z_N)$$

$$= \frac{1}{|\eta(\tau)|^{2\ell}} \sum_{(r,s)\in\Lambda_L} \left(q^{\frac{1}{2}r^2} \bar{q}^{\frac{1}{2}s^2} z_1^{<\alpha_1,r>} \bar{z}_1^{<\alpha_1,s>} \cdots z_N^{<\alpha_N,r>} \bar{z}_N^{<\alpha_N,s>} \right)$$
$$\cdot \prod_{i<j} \left| \chi_\tau(z_i, z_j) \right|^{2<\alpha_i,\alpha_j>}.$$

(3) We have

$$\widetilde{Z}_{E,L}(e^{(\alpha_1,\alpha_1)}, z_1) \cdots (e^{(\alpha_N,\alpha_N)}, z_N)$$

$$= \frac{1}{|\eta(\tau)|^{2\ell}} \sum_{(r,s)\in\Omega_L} \left(q^{\frac{1}{2}r^2} \bar{q}^{\frac{1}{2}s^2} z_1^{<\alpha_1,r>} \bar{z}_1^{<\alpha_1,s>} \cdots z_N^{<\alpha_N,r>} \bar{z}_N^{<\alpha_N,s>} \right)$$
$$\cdot \prod_{i<j} \left| \chi_\tau(z_i, z_j) \right|^{2<\alpha_i,\alpha_j>}$$

$$= \mathrm{Tr}_W \, \widetilde{Y}(e^{(\alpha_1,\alpha_1)}, z_1) \cdots \widetilde{Y}(e^{(\alpha_N,\alpha_N)}, z_N) \cdot q^d \bar{q}^{\bar{d}}.$$

Remark. This is the first indication of the possibility of obtaining the action of vertex operators by means of string path integrals. We will carry this out in the next section.

Proof : The equation (1) is immediate from the definition. The computations of harmonic parts are done in Proposition (A.3), and using them, it is easy to prove (2) and (3).

§9-B. String Path Integrals on Cylinders.

DEFINITION (9.7).
Let $u_1, \ldots, u_N$ be points on $C = C_{\tau',\tau''}$ such that $|z_1| > \cdots > |z_N|$ where $z_j = e^{iu_j}$. Let S_j be the circle given by $\theta \mapsto \theta + u_j$ so that $\theta = 0$ on S_j corresponds to $z = z_j$. Let $\alpha_1, \ldots, \alpha_N \in L'$.

(1) We define

$$K_C(e^{(\alpha_1,\alpha_1)}, z_1) \cdots (e^{(\alpha_N,\alpha_N)}, z_N)$$

$$= \lim_{r\uparrow 1} c \cdot \int_{(C^\infty(\mathbf{S}^1,\mathfrak{h}_{\mathbf{R}})^\wedge)^N} {}^\bullet_\bullet e^{i<\alpha_1 g_r, f_1>} {}^\bullet_\bullet \cdots {}^\bullet_\bullet e^{i<\alpha_N g_r, f_N>} {}^\bullet_\bullet [e^{-I_C(\Phi(f_1,\ldots,f_N))} df_1 \cdots df_N],$$

where $\Phi = \Phi(f_1, \ldots, f_N)$ is the unique function on C such that $\Phi|_{S_i} = f_i$ and $\Phi|_{\partial C} = 0$ which is harmonic on C except on the circles $S_1, \ldots, S_N$. $g_r(\theta) = \sum_{n\in\mathbf{Z}} r^{|n|} e^{in\theta}$ and so g_r approaches the delta function

$$\delta_0(\theta) = \sum_{n\in\mathbf{Z}} e^{in\theta}.$$

c is a normalization constant,

$$c = K_C \Big/ \int_{(C^\infty(\mathbf{S}^1, \mathfrak{h}_{\mathbb{R}})^\wedge)^N} [e^{-I_C(\Phi(f_1,\ldots,f_N)}df_1\cdots df_N].$$

(See Definition (4.6) for the definition of Wick ordering.)

Since

$$\lim_{r\uparrow 1} <\alpha_j g_r, f_j> = <\alpha_j, f_j(0)> = <\alpha_j, \Phi(z_j)>,$$

we regard $K_C(e^{(\alpha_1,\alpha_1)}, z_1)\cdots(e^{(\alpha_N,\alpha_N)}, z_N)$ as a string path integral symbolically written as

$$\int_{C^\infty{}_C(0,0)} {}_\bullet^\bullet e^{i<\alpha_1,\phi(z_1)>}{}_\bullet^\bullet \cdots {}_\bullet^\bullet e^{i<\alpha_N,\phi(z_N)>}{}_\bullet^\bullet e^{-I(\phi)}[d\phi].$$

For $f, g \in C^\infty(\mathbf{S}^1, \mathfrak{h}_{\mathbb{R}})$, we define

$$K_C(e^{(\alpha_1,\alpha_1)}, z_1)\cdots(e^{(\alpha_N,\alpha_N)}, z_N)(f,g)$$
$$= K_C(e^{(\alpha_1,\alpha_1)}, z_1)\cdots(e^{(\alpha_N,\alpha_N)}, z_N) \cdot e^{i<\alpha_1,\phi_0(z_1)>}\ldots e^{i<\alpha_N,\phi_0(z_N)>} \cdot e^{-I(\phi_0)},$$

where $\phi_0 = \phi_C(f,g)$. We regard $K_C(e^{(\alpha_1,\alpha_1)}, z_1)\cdots(e^{(\alpha_N,\alpha_N)}, z_N)(f,g)$ as a string path integral

$$\int_{C^\infty{}_C(f,g)} {}_\bullet^\bullet e^{i<\alpha_1,\phi(z_1)>}{}_\bullet^\bullet \cdots {}_\bullet^\bullet e^{i<\alpha_N,\phi(z_N)>}{}_\bullet^\bullet e^{-I(\phi)}[d\phi].$$

(2) For $f, g \in C^\infty(\mathbf{S}^1, \mathbf{T}_L)$, we define

$$K_{C,L}(e^{(\alpha_1,\alpha_1)}, z_1)\cdots(e^{(\alpha_N,\alpha_N)}, z_N)(f,g)$$
$$= K_C(e^{(\alpha_1,\alpha_1)}, z_1)\cdots(e^{(\alpha_N,\alpha_N)}, z_N) \sum_{\phi_0} e^{i<\alpha_1,\phi_0(z_1)>}\ldots e^{i<\alpha_N,\phi_0(z_N)>} \cdot e^{-I(\phi_0)},$$

where the sum is over all harmonic maps $\phi_0 \in C^\infty{}_{C,L}(f,g)$.

We regard $K_{C,L}(e^{(\alpha_1,\alpha_1)}, z_1)\cdots(e^{(\alpha_N,\alpha_N)}, z_N)(f,g)$ as a string path integral

$$\int_{C^\infty{}_{C,L}(f,g)} {}_\bullet^\bullet e^{i<\alpha_1,\phi(z_1)>}{}_\bullet^\bullet \cdots {}_\bullet^\bullet e^{i<\alpha_N,\phi(z_N)>}{}_\bullet^\bullet e^{-I(\phi)}[d\phi].$$

(3) For $f, g \in \widetilde{C^\infty}(\mathbf{S}^1, \mathbf{T}_L)$, we define

$$\widetilde{K}_{C,L}(e^{(\alpha_1,\alpha_1)}, z_1)\cdots(e^{(\alpha_N,\alpha_N)}, z_N)(f,g)$$
$$= K_C(e^{(\alpha_1,\alpha_1)}, z_1)\cdots(e^{(\alpha_N,\alpha_N)}, z_N) \sum_{\phi_0} e^{i<\alpha_1,\phi_0(z_1)>}\ldots e^{i<\alpha_N,\phi_0(z_N)>}(-1)^{\phi_0}e^{-I(\phi_0)},$$

where the sum is over all harmonic maps $\phi_0 \in C^\infty{}_{C,L}(f,g)$.

We regard $\widetilde{K}_{C,L}(e^{(\alpha_1,\alpha_1)}, z_1)\cdots(e^{(\alpha_N,\alpha_N)}, z_N)(f, g)$ as a string path integral

$$\int_{C^\infty{}_{C,L}(f,g)} {}^\bullet_\bullet e^{i<\alpha_1,\phi(z_1)>}{}^\bullet_\bullet \cdots {}^\bullet_\bullet e^{i<\alpha_N,\phi(z_N)>}{}^\bullet_\bullet (-1)^\phi e^{-I(\phi)} [d\phi].$$

DEFINITION (9.8). [Green Function on Cylinder]

Let w be a point on the cylinder $C = C_{\tau',\tau''}$. We define the Green function of C at w by the formula

$$G_w(z) = \lim_{s\to 1} \sum_{n\in\mathbb{Z}, m\geq 1} \frac{1}{\lambda_{n,m}{}^s} \omega_{n,m}(z)\,\omega_{n,m}(w).$$

where the limit means taking the analytic continuation and setting $s = 1$. (See Proposition (8.6) for the definition of $\lambda_{n,m}$ and $\omega_{n,m}$.) Because $G_w(z) = G_z(w)$, we denote $G_w(z)$ also by $G(z, w)$. We have

$$\ll \phi, G_w \gg = \phi(w),$$

for all $\phi \in C^\infty{}_C(0,0)$. Also we have

$$\ll G_z, G_w \gg = G(z, w), \quad \text{for } z \neq w.$$

Also note that $G(z, w)$ is harmonic in both variables z and w in the domain $z \neq w$.

PROPOSITION (9.9).

For $z, w \in C, z \neq w$, we have

$$\ll G_z, G_w \gg = \frac{1}{2\pi}\left\{ -\frac{y'r'}{2\pi\tau_2} - \log\left|\theta_{11}(\frac{u'-v'}{2\pi}, 2i\tau_2)\right| + \log\left|\theta_{11}(\frac{u'-\overline{v'}}{2\pi}, 2i\tau_2)\right| \right\}$$

$$= -\frac{1}{2\pi}\left\{ \frac{y'r'}{2\pi\tau_2} + \log\left|\frac{\chi_{2i\tau_2}(z',w')}{\chi_{2i\tau_2}(z',\frac{1}{\overline{w'}})}\right| \right\},$$

where $z = e^{iu}, w = e^{iv}, u = x + iy, v = p + ir$ and $z' = e^{iu'}, w' = e^{iv'}, u' = u - 2\pi\tau' = x' + iy', v' = v - 2\pi\tau' = p' + ir'$.

Proof : Notice that

$$\ll G_z, G_w \gg$$
$$= \lim_{s\to 1} \sum_{n\in\mathbb{Z}, m\geq 1} \left(\frac{2\tau_2}{4\pi^2|m + n\cdot 2i\tau_2|^2}\right)^s e^{in(x'-p')}$$
$$\cdot \left(e^{im\frac{y'-r'}{2\tau_2}} + e^{-im\frac{y'-r'}{2\tau_2}} - e^{i\frac{y'+r'}{2\tau_2}} - e^{-im\frac{y'+r'}{2\tau_2}} \right)$$

$$= \lim_{s\to 1} \sum_{n\in\mathbb{Z}, m\neq 0} \left(\frac{2\tau_2}{4\pi^2|m + n\cdot 2i\tau_2|^2}\right)^s e^{in(x'-p')} e^{im\frac{y'-r'}{2\tau_2}}$$
$$- \lim_{s\to 1} \sum_{n\in\mathbb{Z}, m\neq 0} \left(\frac{2\tau_2}{4\pi^2|m + n\cdot 2i\tau_2|^2}\right)^s e^{in(x'-p')} e^{-im\frac{y'+r'}{2\tau_2}}$$

$$= \frac{1}{4\pi^2} \zeta_{2i\tau_2}\left(1; \frac{1}{2\pi}(x'-p'), -\frac{1}{2\pi}\frac{y'-r'}{2\tau_2}\right) - \frac{1}{4\pi^2} \zeta_{2i\tau_2}\left(1; \frac{1}{2\pi}(x'-p'), -\frac{1}{2\pi}\frac{y'+r'}{2\tau_2}\right).$$

The last formula can be easily evaluated using the Kronecker's second limit formula (Lemma (9.3)) and we get

$$= \frac{1}{2\pi}\left\{\frac{(y'-r')^2}{8\pi\tau_2} - \log\left|\theta_{11}\left(\frac{u'-v'}{2\pi}, 2i\tau_2\right)\right| - \frac{(y'+r')^2}{8\pi\tau_2} + \log\left|\theta_{11}\left(\frac{u'-\overline{v'}}{2\pi}, 2i\tau_2\right)\right|\right\}.$$

DEFINITION **(9.10)**.

We define the non-singular part of $\ll G_z, G_w \gg$ by the equality

$$\vdots \ll G_z, G_w \gg \vdots \; = \ll G_z, G_w \gg + \frac{1}{2\pi}\log\left|\frac{z-w}{\sqrt{zw}}\right|.$$

We can take the limit of $w \to z$, and we define

$$\vdots \ll G_z, G_z \gg \vdots \; = \lim_{w\to z} \vdots \ll G_z, G_w \gg \vdots \; = -\frac{1}{2\pi}\left\{\frac{y'^2}{2\pi\tau_2} - \log\left|\chi_{2i\tau_2}\left(z, \frac{1}{z'}\right)\right|\right\},$$

PROPOSITION **(9.11)**.

Let $u_1, \ldots, u_N$ be points on $C = C_{\tau', \tau''}$ such that $|z_1| > \cdots > |z_N|$ where $z_j = e^{iu_j}$. We have

$$K_C(e^{(\alpha_1, \alpha_1)}, z_1)\cdots(e^{(\alpha_N, \alpha_N)}, z_N)$$

$$= K_C \prod_i \exp(-2\pi \vdots \ll \alpha_i G_{z_i}, \alpha_i G_{z_i} \gg \vdots) \cdot \prod_{i\neq j}\exp(-2\pi \ll \alpha_i G_{z_i}, \alpha_j G_{z_j} \gg)$$

$$= K_C \cdot \exp\left(\frac{1}{2\pi\tau_2}(\alpha_1 y_1' + \cdots + \alpha_N y_N')^2\right)$$

$$\cdot \prod_i \left|\chi_{2i\tau_2}\left(z_i', \frac{1}{\overline{z_i'}}\right)\right|^{-<\alpha_i,\alpha_i>} \cdot \prod_{i\neq j}\left|\frac{\chi_{2i\tau_2}(z_i', z_j')}{\chi_{2i\tau_2}(z_i', \frac{1}{\overline{z_j'}})}\right|^{<\alpha_i,\alpha_j>},$$

where $u_j = x_j + iy_j$, $z_j' = e^{iu_j'}$, $u_j' = u_j - 2\pi\tau' = x_j' + iy_j'$.

Remark.

(1) When we take the limit of $\tau_2'' \to \infty$, we have

$$K_C(e^{(\alpha_1, \alpha_1)}, z_1)\cdots(e^{(\alpha_N, \alpha_N)}, z_N)/K_C \to \prod_i\left|\frac{z_i'}{1 - z_i'\overline{z_i'}}\right|^{<\alpha_i,\alpha_i>} \cdot \prod_{i\neq j}\left|\frac{z_i' - z_j'}{1 - z_i'\overline{z_j'}}\right|^{<\alpha_i,\alpha_j>}.$$

(2) When we take the limit of $\tau_2' \to -\infty$ and $\tau_2'' \to \infty$, we have

$$K_C(e^{(\alpha_1, \alpha_1)}, z_1)\cdots(e^{(\alpha_N, \alpha_N)}, z_N)/K_C \to \;<0|\widetilde{Y}(e^{(\alpha_1, \alpha_1)}, z_1)\cdots\widetilde{Y}(e^{(\alpha_N, \alpha_N)}, z_N)|0>,$$

if $\alpha_1 + \cdots + \alpha_N = 0$.

(3) Also note that we have

$$K_C(e^{(\alpha_1,\alpha_1)}, z_1)\cdots(e^{(\alpha_N,\alpha_N)}, z_N)/K_C$$

$$= \left\{ Z_E(e^{(\alpha_1,\alpha_1)}, z_1)\cdots(e^{(\alpha_N,\alpha_N)}, z_N)(e^{(-\alpha_1,-\alpha_1)}, \overline{z}_1^{-1})\cdots(e^{(-\alpha_N,-\alpha_N)}, \overline{z}_N^{-1})/Z_E \right\}^{\frac{1}{2}},$$

where the elliptic curve $E = E_{2i\tau_2}$ is the double of the cylinder $C = C_{0,\tau}$.

PROPOSITION (9.12).

We can compute the kernel functions we defined in Definition (9.7), and we have

(1) $K_{C,L}(e^{(\alpha_1,\alpha_1)}, z_1)\cdots(e^{(\alpha_N,\alpha_N)}, z_N)(f, g)$

$$= K_C(e^{(\alpha_1,\alpha_1)}, z_1)\cdots(e^{(\alpha_N,\alpha_N)}, z_N)(f_*, g_*) \cdot \exp(-\tfrac{1}{2\pi\tau_2}(\alpha_1 y_1' + \cdot\cdot + \alpha_N y_N')^2)$$

$$\cdot \sum_{(r,s)\in\Lambda_L} \left(q^{\frac{1}{2}r^2}\overline{q}^{\frac{1}{2}s^2} \cdot z_1'^{<\alpha_1,r>}\overline{z_1'}^{<\alpha_1,s>}\cdots z_N'^{<\alpha_N,r>}\overline{z_N'}^{<\alpha_N,s>} \right.$$

$$\left. \cdot e^{(r,s)}(f_0 + \lambda\theta)\cdot \overline{e^{(r,s)}}(g_0 + \mu\theta)\right)\frac{\sqrt{2\tau_2}^{\ell}}{v_L},$$

(2) $\widetilde{K}_{C,L}(e^{(\alpha_1,\alpha_1)}, z_1)\cdots(e^{(\alpha_N,\alpha_N)}, z_N)(f, g)$

$$= K_C(e^{(\alpha_1,\alpha_1)}, z_1)\cdots(e^{(\alpha_N,\alpha_N)}, z_N)(f_*, g_*) \cdot \exp(-\tfrac{1}{2\pi\tau_2}(\alpha_1 y_1' + \cdot\cdot + \alpha_N y_N')^2)$$

$$\cdot \sum_{(r,s)\in\Omega_L} \left(q^{\frac{1}{2}r^2}\overline{q}^{\frac{1}{2}s^2} \cdot z_1'^{<\alpha_1,r>}\overline{z_1'}^{<\alpha_1,s>}\cdots z_N'^{<\alpha_N,r>}\overline{z_N'}^{<\alpha_N,s>} \right.$$

$$\left. \cdot e^{(r,s)}(f_0 + \lambda\theta)\cdot \overline{e^{(r,s)}}(g_0 + \mu\theta)\right)\frac{\sqrt{2\tau_2}^{\ell}}{v_L},$$

where $f = f_* + f_0 + \lambda\theta, g = g_* + g_0 + \mu\theta$.

Proof : First, note that the harmonic map in $C^\infty_{C,L}(f, g)$ does not exist when $\lambda \neq \mu$. Also when $\lambda = \mu$, the harmonic maps in $C^\infty_{C,L}(f, g)$ are written as

$$\phi_0 = \phi_{0*} + f_0 + S_{\mu,\beta+(f_0-g_0)/2\pi},$$

where $\phi_{0*} = \phi_C(f_*, g_*), \beta \in L$, and

$$S_{\mu,\beta+(f_0-g_0)/2\pi}(z) = \mu\left((x - 2\pi\tau_1') - \frac{\tau_1}{\tau_2}(y - 2\pi\tau_2')\right) - \left(\beta + \frac{f_0-g_0}{2\pi}\right)\frac{y - 2\pi\tau_2'}{\tau_2}.$$

Furthermore we have

$$I(\phi_0) = I(\phi_{0*}) + I(S_{\mu,\beta+(f_0-g_0)/2\pi}).$$

Therefore, we get

(1) $K_{C,L}(e^{(\alpha_1,\alpha_1)}, z_1)\cdots(e^{(\alpha_N,\alpha_N)}, z_N)(f, g)$

$$= K_{C,L}(e^{(\alpha_1,\alpha_1)}, z_1)\cdots(e^{(\alpha_N,\alpha_N)}, z_N)(f_*, g_*) \cdot \delta_{\lambda,\mu}$$

$$\cdot \sum_{\beta \in L} \exp(i < \alpha_1, S_{\mu,\beta+(f_0-g_0)/2\pi}(z_1) >)\cdots\exp(i < \alpha_N, S_{\mu,\beta+(f_0-g_0)/2\pi}(z_N) >)$$

$$\cdot e^{-I(S_{\mu,\beta+(f_0-g_0)/2\pi})}.$$

(2) $\widetilde{K}_{C,L}(e^{(\alpha_1,\alpha_1)}, z_1)\cdots(e^{(\alpha_N,\alpha_N)}, z_N)(f,g)$

$$= K_{C,L}(e^{(\alpha_1,\alpha_1)}, z_1)\cdots(e^{(\alpha_N,\alpha_N)}, z_N)(f_*, g_*)\cdot\delta_{\lambda,\mu}$$

$$\cdot\sum_{\beta\in L}\exp(i<\alpha_1, S_{\mu,\beta+(f_0-g_0)/2\pi}(z_1)>)\cdots\exp(i<\alpha_N, S_{\mu,\beta+(f_0-g_0)/2\pi}(z_N)>)$$

$$\cdot(-1)^{<\mu,\beta>}e^{-I(S_{\mu,\beta+(f_0-g_0)/2\pi})}.$$

Using the calculation done in Proposition (A.3), we obtain the formulas.

COROLLARY (9.13).
We have the following transformation properties.

(1) $\widetilde{K}_{C,L}(e^{(\alpha_1,\alpha_1)}, z_1)\cdots(e^{(\alpha_N,\alpha_N)}, z_N)(f+2\pi\nu, g)$

$$= (-1)^{<\lambda,\nu>}\widetilde{K}_{C,L}(e^{(\alpha_1,\alpha_1)}, z_1)\cdots(e^{(\alpha_N,\alpha_N)}, z_N)(f,g),$$

$\widetilde{K}_{C,L}(e^{(\alpha_1,\alpha_1)}, z_1)\cdots(e^{(\alpha_N,\alpha_N)}, z_N)(f, g+2\pi\nu)$

$$= (-1)^{<\mu,\nu>}\widetilde{K}_{C,L}(e^{(\alpha_1,\alpha_1)}, z_1)\cdots(e^{(\alpha_N,\alpha_N)}, z_N)(f,g),$$

(2) $\widetilde{K}_{C,L}(e^{(\alpha_1,\alpha_1)}, z_1)\cdots(e^{(\alpha_N,\alpha_N)}, z_N)(f+2\pi\nu, f+2\pi\nu)$

$$= \widetilde{K}_{C,L}(e^{(\alpha_1,\alpha_1)}, z_1)\cdots(e^{(\alpha_N,\alpha_N)}, z_N)(f,f),$$

for all $\nu\in L$ when $f = f_* + f_0 + \lambda\theta, g = g_* + g_0 + \mu\theta$.

§9-C. Analytic Realization of Vertex Operators.

In this section, we prove that string path integrals on cylinders provide kernel functions for products of the neutral vertex operators.

LEMMA (9.14). [Action of Neutral Vertex Operators]
Let $v\in W$ such that

$$v(f) = P\left(\left\{f_n, \overline{f}_n\right\}_{n=1,\ldots,K}\right)\cdot e^{(r_0,s_0)}(f_0+\lambda\theta),$$

where P is a polynomial function of $f_1,\ldots,f_K,\overline{f}_1,\ldots,\overline{f}_K$, and $(r_0, s_0)\in\Omega_L$, for $f = f_* + f_0 + \lambda\theta$. The action of the Wick ordered product of the neutral vertex operators

$${}^\circ_\circ Y(e^{(\alpha_1,\alpha_1)}, z_1)\cdots Y(e^{(\alpha_N,\alpha_N)}, z_N)^\circ_\circ$$

(See Definition (2.21).) on v is given by

$$\left({}^{\circ}_{\circ} Y(e^{(\alpha_1,\alpha_1)}, z_1) \cdots Y(e^{(\alpha_N,\alpha_N)}, z_N) {}^{\circ}_{\circ} \cdot v \right)(f)$$

$$= \; {}_{\bullet}^{\bullet} \exp\left(i <\alpha_1, \widehat{f}(z_1)> + \cdots + i <\alpha_N, \widehat{f}(z_N)> \right) {}_{\bullet}^{\bullet}$$

$$\cdot z_1^{<\alpha_1,r_0>} \cdots z_N^{<\alpha_N,r_0>} \cdot \overline{z}_1^{<\alpha_1,s_0>} \cdots \overline{z}_N^{<\alpha_N,s_0>}$$

$$\cdot P\left(\left\{ f_n + i \left(\alpha_1 \frac{z_1^{-n} - \overline{z}_1^{\,n}}{n} + \cdots + \alpha_N \frac{z_N^{-n} - \overline{z}_N^{\,n}}{n} \right), \right. \right.$$

$$\left. \left. \overline{f}_n - i \left(\alpha_1 \frac{z_1^{\,n} - \overline{z}_1^{\,-n}}{n} + \cdots + \alpha_N \frac{z_N^{\,n} - \overline{z}_N^{\,-n}}{n} \right) \right\}_{n=1,\ldots,K} \right) \cdot e^{(r_0,s_0)}(f_0 + \lambda\theta),$$

for $f = f_* + f_0 + \lambda\theta$.

Proof : Using the functional realization of the action of Heisenberg algebra (See Proposition (4.24).), we get the statement.

COROLLARY (9.15).
Under the same assumption of Lemma (9.14), the operator

$$q'^{-d}\overline{q'}^{-\overline{d}} \widetilde{Y}(e^{(\alpha_1,\alpha_1)}, z_1) \cdots \widetilde{Y}(e^{(\alpha_N,\alpha_N)}, z_N) q''^{d}\overline{q''}^{\overline{d}}$$

acts on v by

$$\left(q'^{-d}\overline{q'}^{-\overline{d}} \widetilde{Y}(e^{(\alpha_1,\alpha_1)}, z_1) \cdots \widetilde{Y}(e^{(\alpha_N,\alpha_N)}, z_N) q''^{d}\overline{q''}^{\overline{d}} \cdot v \right)(f)$$

$$= \prod_i |z_i'|^{<\alpha_i,\alpha_i>} \cdot \prod_{i<j} |z_i' - z_j'|^{2<\alpha_i,\alpha_j>}$$

$$\; {}_{\bullet}^{\bullet} \exp\left(i <\alpha_1, \widehat{f}(z_1')> + \cdots + i <\alpha_N, \widehat{f}(z_N')> \right) {}_{\bullet}^{\bullet}$$

$$\cdot q^{d}\overline{q}^{\overline{d}} P\left(\left\{ f_n + i \left(\alpha_1 \frac{z_1'^{-n} - \overline{z_1'}^{\,n}}{n} + \cdots + \alpha_N \frac{z_N'^{-n} - \overline{z_N'}^{\,n}}{n} \right), \right. \right.$$

$$\left. \left. \overline{f}_n - i \left(\alpha_1 \frac{z_1'^{\,n} - \overline{z_1'}^{\,-n}}{n} + \cdots + \alpha_N \frac{z_N'^{\,n} - \overline{z_N'}^{\,-n}}{n} \right) \right\}_{n=1,\ldots,K} \right)$$

$$\cdot z_1'^{<\alpha_1,r_0>} \cdots z_N'^{<\alpha_N,r_0>} \cdot \overline{z_1'}^{<\alpha_1,s_0>} \cdots \overline{z_N'}^{<\alpha_N,s_0>} \cdot e^{(r_0,s_0)}(f_0 + \lambda\theta),$$

for $f = f_* + f_0 + \lambda\theta$, where $q^{d}\overline{q}^{\overline{d}}$ is acting on v.

Proof : This is true because we have

$$q'^{-d}\overline{q'}^{-\overline{d}} \widetilde{Y}(e^{(\alpha_1,\alpha_1)}, z_1) \cdots \widetilde{Y}(e^{(\alpha_N,\alpha_N)}, z_N) q''^{d}\overline{q''}^{\overline{d}}$$

$$- V(q'^{-L_0}\overline{q'}^{-\overline{L}_0}, e^{(\alpha_1,\alpha_1)}, z_1') \cdots V(q'^{-L_0}\overline{q'}^{-\overline{L}_0}, e^{(\alpha_N,\alpha_N)}, z_N') q^{d}\overline{q}^{\overline{d}} \cdot \prod_i |z_i|^{<\alpha_i,\alpha_i>}$$

$$= Y(e^{(\alpha_1,\alpha_1)}, z_1') \cdots Y(e^{(\alpha_N,\alpha_N)}, z_N') q^{d}\overline{q}^{\overline{d}} \cdot \prod_i |z_i'|^{<\alpha_i,\alpha_i>}$$

$$= \prod_i |z_i'|^{<\alpha_i,\alpha_i>} \cdot \prod_{i<j} |z_i' - z_j'|^{2<\alpha_i,\alpha_j>} \cdot {}^{\circ}_{\circ} Y(e^{(\alpha_1,\alpha_1)}, z_1') \cdots Y(e^{(\alpha_N,\alpha_N)}, z_N') {}^{\circ}_{\circ} q^{d}\overline{q}^{\overline{d}}.$$

124   HARUO TSUKADA

Theorem **(9.16)**. [String Path Integral Realization of Neutral Vertex Operators]
Under the same assumption of Theorem (9.14), we have the following realization of the action of the products of the neutral vertex operators. Namely,

$$
\int_{C^\infty(\mathbf{S}^1,\mathbf{T}_L)^\wedge} \widetilde{K}_{C,L}(e^{(\alpha_1,\alpha_1)},z_1)\cdots(e^{(\alpha_N,\alpha_N)},z_N)(f,g)v(g)\,e^{-\frac{1}{2}(g_*,\overline{g}_*)}[dg]
$$

$$
= \left(q'^{-d}\overline{q'}^{-\overline{d}}\widetilde{Y}(e^{(\alpha_1,\alpha_1)},z_1)\cdots\widetilde{Y}(e^{(\alpha_N,\alpha_N)},z_N)q''^{d}\overline{q''}^{\overline{d}}\cdot v\right)(f)\,e^{-\frac{1}{2}(f_*,\overline{f}_*)},
$$

for all $v \in W$.

Proof : Let us put $v = v_* \otimes e^{(r_0,s_0)}$ where $v_* \in Sym(\widehat{\mathfrak{h}}^\pm)$ and $(r_0,s_0) \in \Omega_L$. Let us take a generating function

$$
v_* = \prod_{n=1}^{\infty}\sum_{m=0}^{\infty}\frac{1}{m!}(\gamma_n(-n)+\delta_n(n))^m,
$$

where $\gamma_n, \delta_n \in \mathfrak{h}$. Note that

$$
v_*(g_*) = \prod_{n=1}^{\infty}\frac{e^{in<\gamma_n,g_n>}e^{in<\delta_n,\overline{g}_n>}}{e^{-n<\gamma_n,\delta_n>}},
$$

where $g_*(\theta) = \sum_{n\neq 0} g_n e^{in\theta}, g_n \in \mathfrak{h}, g_{-n} = \overline{g}_n$. Then we have

$$
\int_{C^\infty{}_*(\mathbf{S}^1,\mathfrak{h}_{\mathbf{R}})^\wedge} K_C(e^{(\alpha_1,\alpha_1)},z_1)\cdots(e^{(\alpha_N,\alpha_N)},z_N)(f_*,g_*)v_*(g_*)\,e^{-\frac{1}{2}(g_*,\overline{g}_*)}[dg_*d\overline{g}_*]
$$

$$
= K_C(e^{(\alpha_1,\alpha_1)},z_1)\cdots(e^{(\alpha_N,\alpha_N)},z_N)/K_C
$$

$$
\cdot \int_{C^\infty{}_*(\mathbf{S}^1,\mathfrak{h}_{\mathbf{R}})^\wedge} K_C(f_*,g_*)e^{i<\alpha_1,\phi_0(z_1)>}\cdots e^{i<\alpha_N,\phi_0(z_N)>}v_*(g_*)\,e^{-\frac{1}{2}(g_*,\overline{g}_*)}[dg_*d\overline{g}_*].
$$

where $\phi_0 = \phi_C(f_*,g_*)$. Theorem (8.9) implies that

$$\int_{C^\infty_*(\mathbf{S}^1,\mathfrak{h}_{\mathbf{R}})^\wedge} K_C(f_*,g_*) e^{i<\alpha_1,\phi_0(z_1)>}\cdots e^{i<\alpha_N,\phi_0(z_N)>} v_*(g_*) e^{-\frac{1}{2}(g_*,\overline{g}_*)}[dg_* d\overline{g}_*]$$

$$= \int_{C^\infty_*(\mathbf{S}^1,\mathfrak{h}_{\mathbf{R}})^\wedge} K_C(f_*,g_*)$$

$$\cdot \exp\left(i < \alpha_1, \sum_{n=1}^\infty \frac{1}{1-q^n\overline{q}^n}(g_n\overline{q}^n(\overline{z_1'}^{-n} - z_1'^{\,n}) + \overline{g}_n q^n(z_1'^{\,-n} - \overline{z_1'}^{\,n})) > \right)$$

$$\cdots \exp\left(i < \alpha_N, \sum_{n=1}^\infty \frac{1}{1-q^n\overline{q}^n}(g_n\overline{q}^n(\overline{z_N'}^{-n} - z_N'^{\,n}) + \overline{g}_n q^n(z_N'^{\,-n} - \overline{z_N'}^{\,n})) > \right)$$

$$\cdot v_*(g_*) e^{-\frac{1}{2}(g_*,\overline{g}_*)}[dg_* d\overline{g}_*]$$

$$\cdot \exp\left(i < \alpha_1, \sum_{n=1}^\infty \frac{1}{1-q^n\overline{q}^n}(f_n(z_1'^{\,n} - q^n\overline{q}^n\overline{z_1'}^{\,-n}) + \overline{f}_n(\overline{z_1'}^{\,n} - q^n\overline{q}^n z_1'^{\,-n})) > \right)$$

$$\cdots \exp\left(i < \alpha_N, \sum_{n=1}^\infty \frac{1}{1-q^n\overline{q}^n}(f_n(z_N'^{\,n} - q^n\overline{q}^n\overline{z_N'}^{\,-n}) + \overline{f}_n(\overline{z_N'}^{\,n} - q^n\overline{q}^n z_N'^{\,-n})) > \right)$$

$$= \int_{C^\infty_*(\mathbf{S}^1,\mathfrak{h}_{\mathbf{R}})^\wedge} K_C(f_*,g_*)$$

$$\cdot \prod_{n=1}^\infty \exp\left(in < g_n, \frac{1}{n}\frac{\overline{q}^n}{1-q^n\overline{q}^n}(\alpha_1(\overline{z_1'}^{-n} - z_1'^{\,n}) + \cdots + \alpha_N(\overline{z_N'}^{-n} - z_N'^{\,n})) + \gamma_n > \right)$$

$$\cdot \exp\left(in < \overline{g}_n, \frac{1}{n}\frac{q^n}{1-q^n\overline{q}^n}(\alpha_1(z_1'^{\,-n} - \overline{z_1'}^{\,n}) + \cdots + \alpha_N(z_N'^{\,-n} - \overline{z_N'}^{\,n})) + \delta_n > \right)$$

$$\cdot \exp\left(n < \frac{1}{n}\frac{\overline{q}^n}{1-q^n\overline{q}^n}(\alpha_1(\overline{z_1'}^{-n} - z_1'^{\,n}) + \cdots + \alpha_N(\overline{z_N'}^{-n} - z_N'^{\,n})) + \gamma_n, \right.$$
$$\left. \frac{1}{n}\frac{q^n}{1-q^n\overline{q}^n}(\alpha_1(z_1'^{\,-n} - \overline{z_1'}^{\,n}) + \cdots + \alpha_N(z_N'^{\,-n} - \overline{z_N'}^{\,n})) + \delta_n > \right)$$

$$\cdot e^{-\frac{1}{2}(g_*,\overline{g}_*)}[dg_* d\overline{g}_*]$$

$$\cdot \prod_{n=1}^\infty \exp\left(i < f_n, \frac{1}{1-q^n\overline{q}^n}(\alpha_1(z_1'^{\,n} - q^n\overline{q}^n\overline{z_1'}^{\,-n}) + \cdots + \alpha_N(z_N'^{\,n} - q^n\overline{q}^n\overline{z_N'}^{\,-n})) > \right)$$

$$\cdot \exp\left(i < \overline{f}_n, \frac{1}{1-q^n\overline{q}^n}(\alpha_1(\overline{z_1'}^{\,n} - q^n\overline{q}^n z_1'^{\,-n}) + \cdots + \alpha_N(\overline{z_N'}^{\,n} - q^n\overline{q}^n z_N'^{\,-n})) > \right)$$

$$\cdot \exp\left(-n < \frac{1}{n}\frac{\overline{q}^n}{1-q^n\overline{q}^n}(\alpha_1(\overline{z_1'}^{\,-n} - z_1'^{\,n}) + \cdots + \alpha_N(\overline{z_N'}^{\,-n} - z_N'^{\,n})) + \gamma_n, \right.$$
$$\left. \frac{1}{n}\frac{q^n}{1-q^n\overline{q}^n}(\alpha_1(z_1'^{\,-n} - \overline{z_1'}^{\,n}) + \cdots + \alpha_N(z_N'^{\,-n} - \overline{z_N'}^{\,n})) + \delta_n > + n < \gamma_n, \delta_n > \right)$$

$$= \frac{1}{\sqrt{2\tau_2}\ell} \left(q^d \overline{q}^{\overline{d}} \cdot v \right)(f_*) \cdot \exp\left(i < \alpha_1, \widehat{f}(z_1') > + \cdots + i < \alpha_N, \widehat{f}(z_N') > \right)$$

$$\cdot \exp\left(- \sum_{n=1}^\infty \frac{1}{n} \frac{q^n \overline{q}^n}{1 - q^n \overline{q}^n} < (\alpha_1(\overline{z_1'}^{-n} - z_1'^{\,n}) + \cdots + \alpha_N(\overline{z_N'}^{-n} - z_N'^{\,n})), \right.$$

$$\left. (\alpha_1(z_1'^{\,-n} - \overline{z_1'}^{\,n}) + \cdots + \alpha_N(z_N'^{\,-n} - \overline{z_N'}^{\,n})) > \right)$$

$$\cdot \prod_{n=1}^\infty \exp\left(i < \gamma_n, iq^n(\alpha_1(z_1'^{\,-n} - \overline{z_1'}^{\,n}) + \cdots + \alpha_N(z_N'^{\,-n} - \overline{z_N'}^{\,n})) > \right)$$

$$\cdot \exp\left(i < \delta_n, -i\overline{q}^n(\alpha_1(z_1'^{\,n} - \overline{z_1'}^{\,-n}) + \cdots + \alpha_N(z_N'^{\,n} - \overline{z_N'}^{\,-n})) > \right).$$

$$= \frac{1}{\sqrt{2\tau_2}\ell} \left(q^d \overline{q}^{\overline{d}} \cdot v \right)(f_*) \cdot \exp\left(i < \alpha_1, \widehat{f}(z_1') > + \cdots + i < \alpha_N, \widehat{f}(z_N') > \right)$$

$$\cdot \prod_{i,j} \left\{ \prod_{n=1}^\infty \left(1 - q^n \overline{q}^n \frac{z_i'}{z_j'} \right)^{-<\alpha_i,\alpha_j>} \cdot \left(1 - q^n \overline{q}^n \frac{\overline{z_i'}}{\overline{z_j'}} \right)^{-<\alpha_i,\alpha_j>} \right.$$

$$\left. \cdot \left(1 - q^n \overline{q}^n z_i' \overline{z_j'} \right)^{<\alpha_i,\alpha_j>} \cdot \left(1 - q^n \overline{q}^n \frac{1}{z_i' \overline{z_j'}} \right)^{<\alpha_i,\alpha_j>} \right\}$$

$$\cdot \prod_{n=1}^\infty \exp\left(i < \gamma_n, iq^n(\alpha_1(z_1'^{\,-n} - \overline{z_1'}^{\,n}) + \cdots + \alpha_N(z_N'^{\,-n} - \overline{z_N'}^{\,n})) > \right)$$

$$\cdot \exp\left(i < \delta_n, -i\overline{q}^n(\alpha_1(z_1'^{\,n} - \overline{z_1'}^{\,-n}) + \cdots + \alpha_N(z_N'^{\,n} - \overline{z_N'}^{\,-n})) > \right).$$

Therefore if

$$v(f) = P\left(\left\{ f_n, \overline{f}_n \right\}_{n=1,\dots,K} \right) \cdot e^{(r_0,s_0)}(f_0 + \lambda\theta),$$

where P is a polynomial function of $f_1, \dots, f_K, \overline{f}_1, \dots, \overline{f}_K$, and $(r_0, s_0) \in \Omega_L$, for $f = f_* + f_0 + \lambda\theta$. Then we have

$$\int_{C^\infty(\mathbf{S}^1, \mathbf{T}_L)^\wedge} \widetilde{K}_{C,L}(e^{(\alpha_1,\alpha_1)}, z_1) \cdots (e^{(\alpha_N,\alpha_N)}, z_N)(f,g) v(g) \, e^{-\frac{1}{2}(g_*, \overline{g}_*)} [dg]$$

$$= \prod_i \left| \frac{z_i'}{1 - z_i' \overline{z_i'}} \right|^{<\alpha_i,\alpha_i>} \cdot \prod_{i,j} \left| \frac{z_i' - z_j'}{1 - z_i \overline{z_j'}} \right|^{<\alpha_i,\alpha_j>}$$

$$\cdot \exp\left(i < \alpha_1, \widehat{f}(z_1') > + \cdots + i < \alpha_N, \widehat{f}(z_N') > \right)$$

$$q^d \overline{q}^{\overline{d}} P\left(\left\{ f_n + i(\alpha_1 \frac{z_1'^{\,-n} - \overline{z_1'}^{\,n}}{n} + \cdots + \alpha_N \frac{z_N'^{\,-n} - \overline{z_N'}^{\,n}}{n}), \right. \right.$$

$$\left. \left. \overline{f}_n - i(\alpha_1 \frac{z_1'^{\,n} - \overline{z_1'}^{\,-n}}{n} + \cdots + \alpha_N \frac{z_N'^{\,n} - \overline{z_N'}^{\,-n}}{n}) \right\}_{n=1,\dots,K} \right)$$

$$\cdot z_1'^{\,<\alpha_1,r_0>} \cdots z_N'^{\,<\alpha_N,r_0>} \cdot \overline{z_1'}^{\,<\alpha_1,s_0>} \cdots \overline{z_N'}^{\,<\alpha_N,s_0>} \cdot e^{(r,s)}(f_0 + \lambda\theta) e^{-\frac{1}{2}(f_*, \overline{f}_*)},$$

for $f = f_* + f_0 + \lambda\theta$, where $q^d \overline{q}^{\overline{d}}$ is acting on v. This implies the proposition.

Remark. We can also get the string path integral realization of the product of vertex operators

$$\widetilde{Y}(e^{\alpha_1}, z_1) \cdots \widetilde{Y}(e^{\alpha_N}, z_N).$$

To do so, we have to take the holomorphic part of $\widetilde{K}_{C,L}(e^{(\alpha_1,\alpha_1)}, z_1) \cdots (e^{(\alpha_N,\alpha_N)}, z_N)(f,g)$.

§9-D. Sewing of String Path Integrals on Cylinder.

In this section, we prove the sewing property of the string path integrals.

PROPOSITION (9.17).

Let $\tau' = \tau_1' + i\tau_2'$ and $\tau'' = \tau_1'' + i\tau_2''$ be two complex numbers such that $\tau_2'' > \tau_2'$. We put $\tau = \tau'' - \tau'$. Let us take the cylinder $C = C_{\tau',\tau''}$ and the elliptic curve $E = E_\tau$. Let $u_1, \ldots, u_N$ be points on C such that $|z_1| > \cdots > |z_N|$ where $z_j = e^{iu_j}$. Let $\alpha_1, \ldots, \alpha_N \in L'$ such that $\alpha_1 + \cdots + \alpha_N = 0$. Put $z_j' = z_j/q'$. We have

(1)
$$\int_{C^\infty_*(\mathbf{S}^1,\mathfrak{h}_{\mathbf{R}})^\wedge} K_C(e^{(\alpha_1,\alpha_1)}, z_1) \cdots (e^{(\alpha_N,\alpha_N)}, z_N)(f_*, f_*)[df_* d\overline{f}_*] \cdot \sqrt{2}^\ell$$
$$= Z_E(e^{(\alpha_1,\alpha_1)}, z_1') \cdots (e^{(\alpha_N,\alpha_N)}, z_N').$$

(2)
$$\int_{C^\infty(\mathbf{S}^1,\mathbf{T}_L)^\wedge} K_{C,L}(e^{(\alpha_1,\alpha_1)}, z_1) \cdots (e^{(\alpha_N,\alpha_N)}, z_N)(f,f)[df]$$
$$= Z_{E,L}(e^{(\alpha_1,\alpha_1)}, z_1') \cdots (e^{(\alpha_N,\alpha_N)}, z_N').$$

(3)
$$\int_{C^\infty(\mathbf{S}^1,\mathbf{T}_L)^\wedge} \widetilde{K}_{C,L}(e^{(\alpha_1,\alpha_1)}, z_1) \cdots (e^{(\alpha_N,\alpha_N)}, z_N)(f,f)[df]$$
$$= \widetilde{Z}_{E,L}(e^{(\alpha_1,\alpha_1)}, z_1') \cdots (e^{(\alpha_N,\alpha_N)}, z_N')$$

Proof : First note that

$$\int_{C^\infty_*(\mathbf{S}^1,\mathfrak{h}_{\mathbf{R}})^\wedge} K_C(e^{(\alpha_1,\alpha_1)}, z_1) \cdots (e^{(\alpha_N,\alpha_N)}, z_N)(f_*, f_*)[df_* d\overline{f}_*]$$

$$= K_C(e^{(\alpha_1,\alpha_1)}, z_1) \cdots (e^{(\alpha_N,\alpha_N)}, z_N)$$
$$\cdot \int_{C^\infty_*(\mathbf{S}^1,\mathfrak{h}_{\mathbf{R}})^\wedge} e^{i<\alpha_1,\widetilde{f}_*(z_1)>} \cdots e^{i<\alpha_N,\widetilde{f}_*(z_N)>} [e^{-I(\widetilde{f}_*)} df_* d\overline{f}_*].$$

We can compute

$$\int_{C^\infty_*(\mathbf{S}^1,\mathfrak{h}_\mathbb{R})^\wedge} e^{i<\alpha_1,\widetilde{f}_*(z_1)>}\cdots e^{i<\alpha_N,\widetilde{f}_*(z_N)>}[e^{-I(\widetilde{f}_*)}df_*d\overline{f}_*]$$

$$= \int_{C^\infty_*(\mathbf{S}^1,\mathfrak{h}_\mathbb{R})^\wedge} \prod_{n=1}^{\infty} \exp\left(i<f_n, \frac{1-\overline{q}^n}{1-q^n\overline{q}^n}(\alpha_1 z_1'^{\,n}+\cdots+\alpha_N z_N'^{\,n})\right.$$

$$\left.+\frac{\overline{q}^n(1-q^n)}{1-q^n\overline{q}^n}(\alpha_1\overline{z_1'}^{-n}+\cdots+\alpha_N\overline{z_N'}^{-n})>\right)$$

$$\cdot\exp\left(i<\overline{f}_n, \frac{1-q^n}{1-q^n\overline{q}^n}(\alpha_1\overline{z_1'}^{\,n}+\cdots+\alpha_N\overline{z_N'}^{\,n})\right.$$

$$\left.+\frac{q^n(1-\overline{q}^n)}{1-q^n\overline{q}^n}(\alpha_1 z_1'^{-n}+\cdots+\alpha_N z_N'^{-n})>\right)$$

$$\left[\exp\left(-\sum_{n=1}^{\infty}n\frac{(1-q^n)(1-\overline{q}^n)}{1-q^n\overline{q}^n}f_n\overline{f}_n\right)df_*d\overline{f}_*\right]$$

$$= \int_{C^\infty_*(\mathbf{S}^1,\mathfrak{h}_\mathbb{R})^\wedge} [e^{-I(\widetilde{f}_*)}df_*d\overline{f}_*]\cdot\prod_{i,j}\left(\chi_\tau(z_i',z_j')\cdot\frac{\chi_{2i\tau_2}(z_i',\frac{1}{\overline{z_i'}})}{\chi_{2i\tau_2}(z_i',z_j')}\right)^{<\alpha_i,\alpha_j>}.$$

This implies the formula (1). Using this it is easy to prove (2) and (3).

PROPOSITION (9.18).
Let $\tau''=\tau_1''+i\tau_2''$, $\tau'''=\tau_1'''+i\tau_2'''$, and $\tau''''=\tau_1''''+i\tau_2''''$ be three complex numbers such that $\tau_2''''>\tau_2'''>\tau_2''$. Let us put $\tau=\tau'''-\tau''$ and $\tau'=\tau''''-\tau'''$. Let $C_1=C_{\tau'',\tau'''}$ and $C_2=C_{\tau''',\tau''''}$ be two cylinders, and let $C_3=C_{\tau'',\tau''''}$ be the cylinder obtained by sewing the two cylinders C_1 and C_2. Let $u_1,\ldots,u_N\in C_1$ and $v_1,\ldots,v_M\in C_2$ such that $|z_1|>\cdots>|z_N|>|w_1|>\cdots>|w_M|$ where $z_j=e^{iu_j}$ and $w_j=e^{iv_j}$. Let $\alpha_1,\ldots,\alpha_N,\beta_1,\ldots,\beta_M\in L'$. We have

$$(1)\quad \int_{C^\infty(\mathbf{S}^1,\mathbf{T}_L)^\wedge} K_{C_1,L}(e^{(\alpha_1,\alpha_1)},z_1)\cdots(e^{(\alpha_N,\alpha_N)},z_N)(f,g)$$

$$\cdot K_{C_2,L}(e^{(\beta_1,\beta_1)},w_1)\cdots(e^{(\beta_M,\beta_M)},w_M)(g,h)[dg]$$

$$= K_{C_3,L}(e^{(\alpha_1,\alpha_1)},z_1)\cdots(e^{(\alpha_N,\alpha_N)},z_N)(e^{(\beta_1,\beta_1)},w_1)\cdots(e^{(\beta_M,\beta_M)},w_M)(f,h),$$

for $f,h\in C^\infty(\mathbf{S}^1,\mathbf{T}_L)$.

$$(2)\quad \int_{C^\infty(\mathbf{S}^1,\mathbf{T}_L)^\wedge} \widetilde{K}_{C_1,L}(e^{(\alpha_1,\alpha_1)},z_1)\cdots(e^{(\alpha_N,\alpha_N)},z_N)(f,g)$$

$$\cdot \widetilde{K}_{C_2,L}(e^{(\beta_1,\beta_1)},w_1)\cdots(e^{(\beta_M,\beta_M)},w_M)(g,h)[dg]$$

$$= \widetilde{K}_{C_3,L}(e^{(\alpha_1,\alpha_1)},z_1)\cdots(e^{(\alpha_N,\alpha_N)},z_N)(e^{(\beta_1,\beta_1)},w_1)\cdots(e^{(\beta_M,\beta_M)},w_M)(f,h),$$

for $f,h\in \widetilde{C^\infty}(\mathbf{S}^1,\mathbf{T}_L)$.

Proof : Let us put $f = f_* + f_0 + \lambda\theta, g = g_* + g_0 + \mu\theta, h = h_* + h_0 + \nu\theta$. First, we will compute the integral

$$(*) \qquad \int_{C^\infty_*(\mathbf{S}^1, \mathfrak{h}_{\mathbb{R}})^\wedge} K_{C_1, L}(e^{(\alpha_1, \alpha_1)}, z_1) \cdots (e^{(\alpha_N, \alpha_N)}, z_N)(f_*, g_*)$$

$$\cdot K_{C_2, L}(e^{(\beta_1, \beta_1)}, w_1) \cdots (e^{(\beta_M, \beta_M)}, w_M)(g_*, h_*)[dg_* d\bar{g}_*].$$

Let us put $\phi_0 = \phi_{C_1}(f_*, g_*), \psi_0 = \phi_{C_2}(g_*, h_*)$. Then the above integral $(*)$ is equal to

$$K_{C_1}(e^{(\alpha_1, \alpha_1)}, z_1) \cdots (e^{(\alpha_N, \alpha_N)}, z_N) K_{C_2}(e^{(\beta_1, \beta_1)}, w_1) \cdots (e^{(\beta_M, \beta_M)}, w_M)$$

$$\cdot \int_{C^\infty_*(\mathbf{S}^1, \mathfrak{h}_{\mathbb{R}})^\wedge} e^{i<\alpha_1, \phi_0(z_1)>} \cdots e^{i<\alpha_N, \phi_0(z_N)>}$$

$$\cdot e^{i<\beta_1, \psi_0(w_1)>} \cdots e^{i<\beta_M, \psi_0(w_M)>} \cdot e^{-I(\phi_0) - I(\psi_0)}[dg_* d\bar{g}_*].$$

Let us put $\varphi_0 = \phi_{C_3}(f_*, h_*)$ and also let us define a function $k_* \in C^\infty_*(\mathbf{S}^1, \mathfrak{h}_{\mathbb{R}})$ by

$$k_*(\theta) = \varphi_0(q''' \cdot e^{i\theta}).$$

Then we have

$$\phi_0 = \varphi_0 + \widetilde{\widetilde{g_* - k_*}}, \quad \text{on } C_1,$$

$$\psi_0 = \varphi_0 + \widetilde{\widetilde{g_* - k_*}}, \quad \text{on } C_2,$$

(See Proposition (8.10) for the definition of $\widetilde{\widetilde{g}}$.) and furthermore,

$$I_{C_1}(\phi_0) + I_{C_2}(\psi_0) = I_{C_3}(\varphi_0) + I_{C_3}(\widetilde{\widetilde{g_* - k_*}}).$$

(See the proof of Proposition (6.6).) Therefore the integral $(*)$ is equal to

$$K_{C_1}(e^{(\alpha_1, \alpha_1)}, z_1) \cdots (e^{(\alpha_N, \alpha_N)}, z_N) K_{C_2}(e^{(\beta_1, \beta_1)}, w_1) \cdots (e^{(\beta_M, \beta_M)}, w_M)$$

$$\cdot \int_{C^\infty_*(\mathbf{S}^1, \mathfrak{h}_{\mathbb{R}})^\wedge} e^{i<\alpha_1, \varphi_0(z_1)>} \cdots e^{i<\alpha_N, \varphi_0(z_N)>} \cdot e^{i<\alpha_1, (\widetilde{\widetilde{g_* - k_*}})(z_1)>} \cdots e^{i<\alpha_N, (\widetilde{\widetilde{g_* - k_*}})(z_N)>}$$

$$\cdot e^{i<\beta_1, \varphi_0(w_1)>} \cdots e^{i<\beta_M, \varphi_0(w_M)>} \cdot e^{i<\beta_1, (\widetilde{\widetilde{g_* - k_*}})(w_1)>} \cdots e^{i<\beta_M, (\widetilde{\widetilde{g_* - k_*}})(w_M)>}$$

$$\cdot e^{-I(\varphi_0) - I(\widetilde{\widetilde{g_* - k_*}})}[dg_* d\bar{g}_*]$$

$$= K_{C_1}(e^{(\alpha_1, \alpha_1)}, z_1) \cdots (e^{(\alpha_N, \alpha_N)}, z_N) K_{C_2}(e^{(\beta_1, \beta_1)}, w_1) \cdots (e^{(\beta_M, \beta_M)}, w_M)$$

$$\cdot \int_{C^\infty_*(\mathbf{S}^1, \mathfrak{h}_{\mathbb{R}})^\wedge} e^{i<\alpha_1, \widetilde{\widetilde{g}}_*(z_1)>} \cdots e^{i<\alpha_N, \widetilde{\widetilde{g}}_*(z_N)>}$$

$$\cdot e^{i<\beta_1, \widetilde{\widetilde{g}}_*(w_1)>} \cdots e^{i<\beta_M, \widetilde{\widetilde{g}}_*(w_M)>}[e^{-I(\widetilde{\widetilde{g}}_*)} dg_* d\bar{g}_*]$$

$$\cdot e^{i<\alpha_1, \varphi_0(z_1)>} \cdots e^{i<\alpha_N, \varphi_0(z_N)>} \cdot e^{i<\beta_1, \varphi_0(w_1)>} \cdots e^{i<\beta_M, \varphi_0(w_M)>} \cdot e^{-I(\varphi_0)}.$$

We can explicitly calculate the integration over $C^\infty{}_*(\mathbf{S}^1, \mathfrak{h}_{\mathbb{R}})^\wedge$. We use the notation $z''_j = z_j/q''$ and $w'''_j = w_j/q'''$.

$$\int_{C^\infty{}_*(\mathbf{S}^1,\mathfrak{h}_{\mathbb{R}})^\wedge} e^{i<\alpha_1, \widetilde{\overline{g}}_*(z_1)>} \cdots e^{i<\alpha_N, \widetilde{\overline{g}}_*(z_N)>}$$
$$\cdot e^{i<\beta_1, \widetilde{\overline{g}}_*(w_1)>} \cdots e^{i<\beta_M, \widetilde{\overline{g}}_*(w_M)>} [e^{-I(\widetilde{\overline{g}}_*)} dg_* d\overline{g}_*]$$

$$= \int_{C^\infty{}_*(\mathbf{S}^1,\mathfrak{h}_{\mathbb{R}})^\wedge} \exp\Big(\sum_{n=1}^\infty i < g_n, \frac{\overline{q}^n}{1-q^n\overline{q}^n}(\alpha_1(\overline{z''_1}^{-n} - z''_1{}^n) + \cdots + \alpha_N(\overline{z''_N}^{-n} - z''_N{}^n))$$
$$+ \frac{1}{1-q'^n\overline{q}'^n}(\beta_1(w'''_1{}^n - q'^n\overline{q}'^n\overline{w'''_1}^{-n}) + \cdots + \beta_M(w'''_M{}^n - q'^n\overline{q}'^n\overline{w'''_N}^{-n})) > \Big)$$

$$\cdot \exp\Big(\sum_{n=1}^\infty i < \overline{g}_n, \frac{q^n}{1-q^n\overline{q}^n}(\alpha_1(z''_1{}^{-n} - \overline{z''_1}^n) + \cdots + \alpha_N(z''_N{}^{-n} - \overline{z''_N}^n))$$
$$+ \frac{1}{1-q'^n\overline{q}'^n}(\beta_1(\overline{w'''_1}^n - q'^n\overline{q}'^n w'''_1{}^{-n}) + \cdots + \beta_M(\overline{w'''_M}^n - q'^n\overline{q}'^n w'''_M{}^{-n})) > \Big)$$

$$\cdot \Big[\exp\Big(-\sum_{n=1}^\infty n \frac{1-q^n q'^n \overline{q}^n \overline{q}'^n}{(1-q^n\overline{q}^n)(1-q'^n\overline{q}'^n)} g_n \overline{g}_n \Big) dg_* d\overline{g}_* \Big]$$

$$= \int_{C^\infty{}_*(\mathbf{S}^1,\mathfrak{h}_{\mathbb{R}})^\wedge} [e^{-I(\widetilde{\overline{g}}_*)} dg_* d\overline{g}_*] \cdot \prod_{i,j} \left| \frac{\chi_{2i\tau_2}(z''_i, \frac{1}{\overline{z''_j}})}{\chi_{2i\tau_2}(z''_i, z''_j)} \cdot \frac{\chi_{2i(\tau_2+\tau'_2)}(z''_i, z''_j)}{\chi_{2i(\tau_2+\tau'_2)}(z''_i, \frac{1}{\overline{z''_j}})} \right|^{<\alpha_i,\alpha_j>}$$

$$\cdot \prod_{i,j} \left| \frac{\chi_{2i\tau'_2}(w'''_i, \frac{1}{\overline{w'''_j}})}{\chi_{2i\tau'_2}(w'''_i, w'''_j)} \cdot \frac{\chi_{2i(\tau_2+\tau'_2)}(w''_i, w''_j)}{\chi_{2i(\tau_2+\tau'_2)}(w''_i, \frac{1}{\overline{w''_j}})} \cdot \frac{1}{q} \right|^{<\beta_i,\beta_j>}$$

$$\cdot \prod_{i,j} \left| \frac{\chi_{2i(\tau_2+\tau'_2)}(z''_i, w''_j)}{\chi_{2i(\tau_2+\tau'_2)}(z''_i, \frac{1}{\overline{w''_j}})} \cdot \frac{1}{z''_i} \right|^{2<\alpha_i,\beta_j>}$$

We know from the proof of Proposition (8.10),

$$K_{C_1} K_{C_2} \int_{C^\infty{}_*(\mathbf{S}^1,\mathfrak{h}_{\mathbb{R}})^\wedge} [e^{-I(\widetilde{\overline{g}}_*)} dg_* d\overline{g}_*] = K_{C_3} \sqrt{\frac{\tau_2+\tau'_2}{2\tau_2\tau'_2}}^\ell .$$

Therefore, using Proposition (9.11), we find that the integral (*) is equal to

$$\sqrt{\frac{\tau_2+\tau'_2}{\tau_2\tau'_2}}^\ell \cdot \exp\Big(\frac{1}{2\pi\tau_2}(\alpha_1 y''_1 + \cdots + \alpha_N y''_N)^2 \Big) \exp\Big(\frac{1}{2\pi\tau'_2}(\beta_1 r'''_1 + \cdots + \beta_N r'''_N)^2 \Big)$$

$$\cdot \exp\Big(-\frac{1}{2\pi(\tau_2+\tau'_2)}(\alpha_1 y''_1 + \cdots + \alpha_N y''_N + \beta_1 r''_1 + \cdots + \beta_M r''_M)^2 \Big) \cdot |q|^{-<\beta,\beta>} \cdot \prod_i |z''_i|^{-2<\alpha_i,\beta>}$$

$$\cdot K_{C_3}\big(e^{(\alpha_1,\alpha_1)}, z_1\big) \cdots \big(e^{(\alpha_N,\alpha_N)}, z_N\big)\big(e^{(\beta_1,\beta_1)}, w_1\big) \cdots \big(e^{(\beta_M,\beta_M)}, w_M\big)(0,0)$$

$$\cdot e^{i<\alpha_1, \varphi_0(z_1)>} \cdots e^{i<\alpha_N, \varphi_0(z_N)>} \cdot e^{i<\beta_1, \varphi_0(w_1)>} \cdots e^{i<\beta_M, \varphi_0(w_M)>} \cdot e^{-I(\varphi_0)}$$

$$= \sqrt{\frac{\tau_2 + \tau_2'}{\tau_2 \tau_2'}}^{\ell} \cdot \exp\left(\frac{1}{2\pi\tau_2}(\alpha_1 y_1'' + \cdots + \alpha_N y_N'')^2\right) \exp\left(\frac{1}{2\pi\tau_2'}(\beta_1 r_1''' + \cdots + \beta_N r_N''')^2\right)$$

$$\cdot \exp\left(-\frac{1}{2\pi(\tau_2+\tau_2')}(\alpha_1 y_1'' + \cdots + \alpha_N y_N'' + \beta_1 r_1'' + \cdots + \beta_M r_M'')^2\right) \cdot |q|^{-<\beta,\beta>} \cdot \prod_i |z_i''|^{-2<\alpha_i,\beta>}$$

$$\cdot K_{C_3}(e^{(\alpha_1,\alpha_1)}, z_1) \cdots (e^{(\alpha_N,\alpha_N)}, z_N)(e^{(\beta_1,\beta_1)}, w_1) \cdots (e^{(\beta_M,\beta_M)}, w_M)(f_*, h_*).$$

where $u_j'' = u_j - 2\pi\tau''$, $v_j''' = v_j - 2\pi\tau'''$, $u = x+iy$, $v = p+ir$. Combining this calculation with Proposition (9.12), we get

$$(1) \quad \int_{C^\infty(\mathbf{S}^1, \mathbf{T}_L)^\wedge} K_{C_1,L}(e^{(\alpha_1,\alpha_1)}, z_1) \cdots (e^{(\alpha_N,\alpha_N)}, z_N)(f, g)$$

$$\cdot K_{C_2,L}(e^{(\beta_1,\beta_1)}, w_1) \cdots (e^{(\beta_M,\beta_M)}, w_M)(g, h)[dg]$$

$$= \int_{C^\infty(\mathbf{S}^1, \mathbf{T}_L)^\wedge} K_{C_1}(e^{(\alpha_1,\alpha_1)}, z_1) \cdots (e^{(\alpha_N,\alpha_N)}, z_N)(f_*, g_*) \cdot \exp\left(-\frac{1}{2\pi\tau_2}(\alpha_1 y_1'' + \cdots + \alpha_N y_N'')^2\right)$$

$$\cdot \sum_{(r,s)\in\Lambda_L} \left(q^{\frac{1}{2}r^2} \overline{q}^{\frac{1}{2}s^2} \cdot z_1''^{<\alpha_1,r>} \overline{z''}_1^{<\alpha_1,s>} \cdots z_N''^{<\alpha_N,r>} \overline{z''}_N^{<\alpha_N,s>} \right.$$

$$\left. \cdot e^{(r,s)}(f_0 + \lambda\theta) \cdot \overline{e^{(r,s)}}(g_0 + \mu\theta)\right) \frac{\sqrt{2\tau_2}^{\ell}}{v_L}$$

$$\cdot K_{C_2}(e^{(\beta_1,\beta_1)}, w_1) \cdots (e^{(\beta_M,\beta_M)}, w_M)(g_*, h_*) \cdot \exp\left(-\frac{1}{2\pi\tau_2'}(\beta_1 r_1''' + \cdots + \beta_N r_N''')^2\right)$$

$$\cdot \sum_{(r,s)\in\Lambda_L} \left(q'^{\frac{1}{2}r^2} \overline{q'}^{\frac{1}{2}s^2} \cdot w_1'''^{<\beta_1,r>} \overline{w'''}_1^{<\beta_1,s>} \cdots w_M'''^{<\beta_M,r>} \overline{w'''}_M^{<\beta_M,s>} \right.$$

$$\left. \cdot e^{(r,s)}(g_0 + \mu\theta) \cdot \overline{e^{(r,s)}}(h_0 + \nu\theta)\right) \frac{\sqrt{2\tau_2'}^{\ell}}{v_L} [dg]$$

$$= K_{C_3}(e^{(\alpha_1,\alpha_1)}, z_1) \cdots (e^{(\alpha_N,\alpha_N)}, z_N)(e^{(\beta_1,\beta_1)}, w_1) \cdots (e^{(\beta_M,\beta_M)}, w_M)(f_*, h_*)$$

$$\cdot \exp\left(-\frac{1}{2\pi(\tau_2+\tau_2')}(\alpha_1 y_1'' + \cdots + \alpha_N y_N'' + \beta_1 r_1'' + \cdots + \beta_M r_M'')^2\right) \cdot |q|^{-<\beta,\beta>} \cdot \prod_i |z_i''|^{-2<\alpha_i,\beta>}$$

$$\cdot \sum_{(r,s)\in\Lambda_L} \left(q^{\frac{1}{2}(r+\beta)^2} q'^{\frac{1}{2}r^2} \overline{q}^{\frac{1}{2}(s+\beta)^2} \overline{q'}^{\frac{1}{2}s^2} \right.$$

$$\cdot z_1''^{<\alpha_1,r+\beta>} \overline{z''}_1^{<\alpha_1,s+\beta>} \cdots z_N''^{<\alpha_N,r+\beta>} \overline{z''}_N^{<\alpha_N,r+\beta>}$$

$$\cdot w_1'''^{<\beta_1,r>} \overline{w'''}_1^{<\beta_1,s>} \cdots w_M'''^{<\beta_M,r>} \overline{w'''}_M^{<\beta_M,s>}$$

$$\left. \cdot e^{(r,s)}(f_0 + \lambda\theta) \cdot \overline{e^{(r,s)}}(h_0 + \nu\theta)\right) \frac{\sqrt{2(\tau_2 + \tau_2')}^{\ell}}{v_L}$$

$$= K_{C_3,L}(e^{(\alpha_1,\alpha_1)}, z_1) \cdots (e^{(\alpha_N,\alpha_N)}, z_N)(e^{(\beta_1,\beta_1)}, w_1) \cdots (e^{(\beta_M,\beta_M)}, w_M)(f, h).$$

(2) is proved in a similar way.

APPENDIX

§A. Poisson Summation Formula.

In this appendix, we state the Poisson summation formula and by using it, we carry out several calculations on the harmonic part of the string path integrals in sections 6 and 9.

THEOREM (A.1). [Poisson Summation Formula] (See Serre [Ser].)

Let V be a $\mathbb{R}$-vector space with an inner product $< , >$. Let $f: V \to \mathbb{C}$ be a rapidly decreasing smooth function. We define the Fourier transform $\widehat{f}$ of f by the following formula

$$\widehat{f}(y) = \int_V e^{-2\pi i <x,y>} f(x) dx.$$

Then we have an equation for two kinds of summations

$$\sum_{\beta \in L} f(\beta) = \frac{1}{\text{vol}(V/L)} \sum_{\gamma \in L'} \widehat{f}(\gamma),$$

for any lattice L in V. Here, $\text{vol}(V/L) = \int_{V/L} dx$ and L' is the dual lattice of L.

PROPOSITION (A.2).

We have the following equalities.

(1) $\displaystyle \sum_{\beta \in L} f(\beta + a) = \frac{1}{\text{vol}(V/L)} \sum_{\gamma \in L'} e^{2\pi i <a,\gamma>} \widehat{f}(\gamma).$

(2) $\displaystyle \sum_{\beta \in L} e^{-2\pi i <a,\beta>} f(\beta) = \frac{1}{\text{vol}(V/L)} \sum_{\gamma \in L'} \widehat{f}(\gamma + a).$

for a fixed $a \in V$.

PROPOSITION **(A.3)**.

Under the same assumption of Definition (9.5), We have the following equalities on the harmonic parts of the string path integrals.

(1) (a) $\displaystyle\sum_{\beta\in L} e^{-I(S_{\lambda,\beta+(g_0-f_0)/2\pi})} = \sum_{\gamma\in L'} \left(q^{\frac{1}{2}(\gamma+\frac{\lambda}{2})^2} \bar{q}^{\frac{1}{2}(\gamma-\frac{\lambda}{2})^2} \cdot e^{i<\gamma,f_0-g_0>} \right) \frac{\sqrt{2\tau_2}^{\ell}}{v_L}.$

(b) $\displaystyle\sum_{\alpha,\beta\in L} e^{-I(S_{\alpha,\beta})} = \sum_{(r,s)\in\Lambda_L} q^{\frac{1}{2}r^2} \bar{q}^{\frac{1}{2}s^2} \cdot \frac{\sqrt{2\tau_2}^{\ell}}{v_L}.$

(c) $\displaystyle\sum_{\beta\in L} \exp(i<\alpha_1, S_{\lambda,\beta+(g_0-f_0)/2\pi}(z_1)>) \cdots \exp(i<\alpha_N, S_{\lambda,\beta+(g_0-f_0)/2\pi}(z_N)>)$

$$\cdot e^{-I(S_{\lambda,\beta+(g_0-f_0)/2\pi})}$$

$$= \exp\left(-\frac{1}{2\pi\tau_2}(\alpha_1 y_1' + \cdots + \alpha_N y_N')^2\right)$$

$$\cdot \sum_{\gamma\in L'} \left(q^{\frac{1}{2}(\gamma+\frac{\lambda}{2})^2} \bar{q}^{\frac{1}{2}(\gamma-\frac{\lambda}{2})^2} \cdot z_1'^{<\alpha_1,\gamma+\frac{\lambda}{2}>} \cdots z_N'^{<\alpha_N,\gamma+\frac{\lambda}{2}>} \right.$$

$$\left. \cdot \bar{z_1'}^{<\alpha_1,\gamma-\frac{\lambda}{2}>} \cdots \bar{z_N'}^{<\alpha_N,\gamma-\frac{\lambda}{2}>} \cdot e^{i<\gamma,f_0-g_0>} \right) \frac{\sqrt{2\tau_2}^{\ell}}{v_L}.$$

(d) $\displaystyle\sum_{\alpha,\beta\in L} \exp(i<\alpha_1, S_{\alpha,\beta}(z_1)>) \cdots \exp(i<\alpha_N, S_{\alpha,\beta}(z_N)>) \cdot e^{-I(S_{\alpha,\beta})}$

$$= \exp\left(-\frac{1}{2\pi\tau_2}(\alpha_1 y_1' + \cdots + \alpha_N y_N')^2\right)$$

$$\cdot \sum_{(r,s)\in\Lambda_L} q^{\frac{1}{2}r^2} \bar{q}^{\frac{1}{2}s^2} \cdot z_1'^{<\alpha_1,r>} \cdots z_N'^{<\alpha_N,r>} \cdot \bar{z_1'}^{<\alpha_1,s>} \cdots \bar{z_N'}^{<\alpha_N,s>} \cdot \frac{\sqrt{2\tau_2}^{\ell}}{v_L}.$$

(2) (a) $\displaystyle\sum_{\beta\in L} (-1)^{<\lambda,\beta>} e^{-I(S_{\lambda,\beta+(g_0-f_0)/2\pi})} = \sum_{\gamma\in L'} \left(q^{\frac{1}{2}(\gamma+\lambda)^2} \bar{q}^{\frac{1}{2}\gamma^2} \cdot e^{i<\gamma+\frac{\lambda}{2},f_0-g_0>} \right) \frac{\sqrt{2\tau_2}^{\ell}}{v_L}.$

(b) $\displaystyle\sum_{\alpha,\beta\in L} (-1)^{<\alpha,\beta>} e^{-I(S_{\alpha,\beta})} = \sum_{(r,s)\in\Omega_L} q^{\frac{1}{2}r^2} \bar{q}^{\frac{1}{2}s^2} \cdot \frac{\sqrt{2\tau_2}^{\ell}}{v_L}.$

(c) $\displaystyle\sum_{\beta\in L} \exp(i<\alpha_1, S_{\lambda,\beta+(g_0-f_0)/2\pi}(z_1)>) \cdots \exp(i<\alpha_N, S_{\lambda,\beta+(g_0-f_0)/2\pi}(z_N)>)$

$$\cdot (-1)^{<\lambda,\beta>} e^{-I(S_{\lambda,\beta+(g_0-f_0)/2\pi})}.$$

$$= \exp\left(-\frac{1}{2\pi\tau_2}(\alpha_1 y_1' + \cdots + \alpha_N y_N')^2\right)$$

$$\cdot \sum_{\gamma\in L'} \left(q^{\frac{1}{2}(\gamma+\lambda)^2} \bar{q}^{\frac{1}{2}\gamma^2} \cdot z_1'^{<\alpha_1,\gamma+\lambda>} \cdots z_N'^{<\alpha_N,\gamma+\lambda>} \right.$$

$$\left. \cdot \bar{z_1'}^{<\alpha_1,\gamma>} \cdots \bar{z_N'}^{<\alpha_N,\gamma>} \cdot e^{i<\gamma+\frac{\lambda}{2},f_0-g_0>} \right) \frac{\sqrt{2\tau_2}^{\ell}}{v_L}.$$

(d) $\displaystyle\sum_{\alpha,\beta\in L}\exp(i<\alpha_1,S_{\alpha,\beta}(z_1)>)\cdots\exp(i<\alpha_N,S_{\alpha,\beta}(z_N)>)(-1)^{<\alpha,\beta>}e^{-I(S_{\alpha,\beta})}$

$$= \exp\left(-\frac{1}{2\pi\tau_2}(\alpha_1 y_1' + \cdots + \alpha_N y_N')^2\right)$$
$$\cdot \sum_{(r,s)\in\Omega_L} q^{\frac{1}{2}r^2}\overline{q}^{\frac{1}{2}s^2} \cdot z_1'^{<\alpha_1,r>}\cdots z_N'^{<\alpha_N,r>} \cdot \overline{z_1'}^{<\alpha_1,s>}\cdots \overline{z_N'}^{<\alpha_N,s>} \cdot \frac{\sqrt{2\tau_2}^\ell}{v_L}.$$

Proof : Define a function
$$f_\lambda(x) = e^{-I(S_{\lambda,x})} = e^{-\frac{\pi}{2}(\tau_2<\lambda,\lambda>+\frac{1}{\tau_2}<x-\tau_1\lambda,x-\tau_1\lambda>)}.$$

Then its Fourier transform is
$$\frac{\widehat{f_\lambda}(y)}{\text{vol}(\mathfrak{h}_{\mathbb{R}}/L)} = e^{-\frac{\pi}{2}\tau_2<\lambda,\lambda>-2\pi i \tau_1<\lambda,y>-2\pi\tau_2<y.y>} \cdot \frac{\sqrt{2\tau_2}^\ell}{v_L} = q^{\frac{1}{2}(y-\frac{\lambda}{2})^2}\overline{q}^{\frac{1}{2}(y+\frac{\lambda}{2})^2} \cdot \frac{\sqrt{2\tau_2}^\ell}{v_L}.$$

(1) Therefore we get the following.

(a) $\displaystyle\sum_{\beta\in L} e^{-I(S_{\lambda,\beta+(g_0-f_0)/2\pi})} = \sum_{\beta\in L} f_\lambda(\beta+(g_0-f_0)/2\pi)$

$$= \frac{1}{\text{vol}(\mathfrak{h}_{\mathbb{R}}/L)}\sum_{\gamma\in L'} e^{i<\gamma,g_0-f_0>}\widehat{f_\lambda}(\gamma) = \sum_{\gamma\in L'}\left(q^{\frac{1}{2}(\gamma+\frac{\lambda}{2})^2}\overline{q}^{\frac{1}{2}(\gamma-\frac{\lambda}{2})^2} \cdot e^{i<\gamma,f_0-g_0>}\right)\frac{\sqrt{2\tau_2}^\ell}{v_L}.$$

This implies (b).

(c) $\displaystyle\sum_{\beta\in L}\exp(i<\alpha_1,S_{\lambda,\beta+(g_0-f_0)/2\pi}(z_1)>)\cdots\exp(i<\alpha_N,S_{\lambda,\beta+(g_0-f_0)/2\pi}(z_N)>)$

$$\cdot e^{-I(S_{\lambda,\beta+(g_0-f_0)/2\pi})}$$

$$= \exp\left(i<\alpha_1(x_1' - \frac{\tau_1}{\tau_2}y_1') + \cdots + \alpha_N(x_N' - \frac{\tau_1}{\tau_2}y_N'),\lambda>\right)$$
$$\cdot \sum_{\beta\in L}\exp\left(\frac{i}{\tau_2}<\alpha_1 y_1' + \cdots + \alpha_N y_N', \beta+(g_0-f_0)/2\pi>\right)f_\lambda(\beta+(g_0-f_0)/2\pi)$$

$$= \exp\left(i<\alpha_1(x_1' - \frac{\tau_1}{\tau_2}y_1') + \cdots + \alpha_N(x_N' - \frac{\tau_1}{\tau_2}y_N'),\lambda>\right)$$
$$\cdot \frac{1}{\text{vol}(\mathfrak{h}_{\mathbb{R}}/L)}\sum_{\gamma\in L'} e^{i<\gamma,g_0-f_0>}\widehat{f_\lambda}\left(\gamma+\frac{1}{2\pi\tau_2}(\alpha_1 y_1' + \cdots + \alpha_N y_N')\right)$$

$$= \exp\left(-\frac{1}{2\pi\tau_2}(\alpha_1 y_1' + \cdots + \alpha_N y_N')^2\right)$$
$$\cdot \sum_{\gamma\in L'}\left(q^{\frac{1}{2}(\gamma+\frac{\lambda}{2})^2}\overline{q}^{\frac{1}{2}(\gamma-\frac{\lambda}{2})^2} \cdot z_1'^{<\alpha_1,\gamma+\frac{\lambda}{2}>}\overline{z_1'}^{<\alpha_1,\gamma-\frac{\lambda}{2}>}\right.$$

$$\left.\cdots z_N'^{<\alpha_N,\gamma+\frac{\lambda}{2}>}\overline{z_N'}^{<\alpha_N,\gamma-\frac{\lambda}{2}>} \cdot e^{i<\gamma,f_0-g_0>}\right)\frac{\sqrt{2\tau_2}^\ell}{v_L}.$$

This implies (d).

(2) Since in general
$$\sum_{\beta\in L} f(\beta)(-1)^{<\lambda,\beta>} = \frac{1}{\text{vol}(\mathfrak{h}_{\mathbb{R}}/L)}\sum_{\gamma\in L'}\widehat{f}\left(\gamma+\frac{\lambda}{2}\right),$$

we get the equations of (2) immediately from (1).

REFERENCES

[ADDF] M. Ademollo, E. Del Guidice, P. Di Vecchia and S. Fubini: Couplings of Three Excited Particles in the Dual-Resonance Model, *Nuovo Cim.* **19A** (1974) pp.181–203.

[BPZ] A. A. Belavin, A. M. Polyakov and A. B. Zamolodchikov: Infinite Conformal Symmetries in Two-dimensional Quantum Field Theory, *Nucl. Phys.* **B241** (1984) pp.333–380.

[Bor] R. E. Borcherds: Vertex algebras, Kac-Moody algebras and the Monster, *Proc. Natl. Acad. Sci. U.S.A.* **83** (1986) pp.3068–3071.

[Bos] J.-B. Bost: *Conformal and Holomorphic Anomalies on Riemann Surfaces and Determinant Line Bundles*, preprint (1986).

[BT] R. C. Brower and C. B. Thorn: Eliminating Spurious States from the Dual Resonance Model, *Nucl. Phys.* **B31** (1971) pp.163–182.

[CENT] A. Casher, F. Englert, H. Nicolai and A. Taormina: Consistent Superstrings as solutions of the D=26 Bosonic String Theory, *Phys. Lett.* **162B** (1985) pp.121–126.

[DJKM] E. Date, M. Jimbo, M. Kashiwara and T. Miwa: Transformation Groups for Soliton Equations, *Publ. RIMS Kyoto Univ.* **18** (1982) pp.1077–1110.

[DDF] E. Del Guidice, P. Di Vecchia and S. Fubini: General Properties of the Dual-Resonance Model, *Ann. Phys.* **70** (1972) pp.378–398.

[DP] E. D'Hoker and D. H. Phong: Length-Twist Parameters in String Path Integrals, *Phys. Rev. Lett.* **56** (1986) pp.912–915.

[ES] J. Eells and J. H. Sampson: Harmonic Mappings of Riemannian Manifolds, *Amer. J. Math.* **86** (1964) pp.109–160.

[Fo] R. Forman: Functional Determinants and Geometry, *Invent. Math.* **88** (1987) pp.447–493.

[Fr] I. B. Frenkel: Two Constructions of Affine Lie Algebra Representations and Boson-Fermion Correspondence in Quantum Field Theory, *J. Funct. Anal.* **44** (1981) pp.259–327.

[FK] I. B. Frenkel and V. G. Kac: Basic Representations of Affine Lie Algebras and Dual Resonance Models, *Inv. Math.* **62** (1980) pp.23–66.

[FLM1] I. B. Frenkel, J. Lepowsky and A. Meurman: *Vertex Operator Calculus*, in "Mathematical Aspects of String Theory", Ed. by S. T. Yau, World Scientific (1987) pp.150–188.

[FLM2] I. B. Frenkel, J. Lepowsky and A. Meurman: *Vertex Operator Algebras and the Monster*, Academic Press (1988).

[FGV]   S. Fubini, D. Gordon and G. Veneziano: A General Treatment of Factorization
        in Dual Resonance Models, *Phys. Lett.* **29B** (1969) pp.679–682.

[FV]   S. Fubini and G. Veneziano: Algebraic Treatment of Subsidiary Conditions in
        Dual Resonance Models, *Ann. Phys.* **63** (1971) pp.12–27.

[GV]   I. M. Gel'fand and N. Ya. Vilenkin: *Generalized Functions*, Vol 4, *Applications
        of Harmonic Analysis*, (Translated by A. Feinstein) Academic Press (1964).

[GJ]   J. Glimm and A. Jaffe: *Quantum Physics, A Funcional Integral Point of View*,
        Springer (First Edition 1981, Second Edition 1987).

[GNSS] D. J. Gross, A. Neveu, J. Scherk, and J. H. Schwarz: Renormalization and
        Unitarity in the Dual-Resonance Model, *Phys. Rev.* **D2** (1970) pp.697–710.

[GSW]   M. B. Green, J. H. Schwarz and E. Witten: *Superstring Theory*, Vols. 1 and 2,
        Cambridge University Press (1987).

[Ham]   R. S. Hamilton: *Harmonic Maps of Manifolds with Boundary*, Springer Lecture
        Notes in Mathematics no.**471** (1975).

[Haw]   S. W. Hawking: Zeta Function Regularization of Path Integrals IN Curved
        Spacetime, *Comm. Math. Phys.* **55** (1977) pp.133–148.

[Hö]   L. Hörmander: *Linear Partial Differential Operators*, Springer (1963).

[HSV]   C. S. Hsue, B. Sakita and M. A. Virasoro: Formulation of Dual Theory in Terms
        of Functional Integrations, *Phys. Rev.* **D12** (1970) pp. 2857–2868.

[IZ]   C. Itzykson and J-B. Zuber: *Quantum Field Theory*, McGraw-Hill (1980).

[KKLW] V. G. Kac, D. A. Kazhdan, J. Lepowsky, and R. L. Wilson: Realization of the
        Basic Representations of the Euclidean Lie Algebras, *Adv. in Math.* **42** (1981)
        pp.83–112.

[KKW]   S. Klimek, W. Kondracki, and M. Wisniowski: The ζ-determinant and Func-
        tional Analysis, *Bull. Sc. Math.*, 2^e serie **113** (1989) pp.175–193.

[Kr]   L. Kronecker: Zur Theorie der Elliptischen Functionen I, IV (1883) in Leopold
        Kronecker's Werke IV.

[Ku]   H-H. Kuo: *Gaussian Measures in Banach Spaces*, Springer Lecture Notes in
        Mathematics no.**463** (1975).

[LP]   J. Lepowsky and M. Primc: Standard Modules for the Type One Affine Lie
        Algebras, in "Number Theory, New York", Springer Lecture Notes in Mathe-
        matics no.**1052** (1984) pp.194–251.

[LW1]   J. Lepowsky and R. L. Wilson: Construction of the Affine Lie Algebra $A_1^{(1)}$,
        Comm. Math. Phys. **62** (1978) pp.43–53.

[LW2]   J. Lepowsky and R. L. Wilson: A Lie Theoretic Interpretation and Proof of the
        Rogers-Ramanujan Identities, *Advances in Math.* **45** (1982) pp.21–72.

[M]   S. Mandelstam: Dual Resonance Models, *Phys. Reports* **13** (1974) pp.259–353.

[MS]   H. P. McKean, Jr. and I. M. Singer: Curvature and the Eigenvalues of the
        Laplacian, *J. Diff. Geom.* **1** (1967) pp.43–69.

[MP] S. Minakshisundaram and A. Pleijel: Some Properties of the Eigenfunctions of the Laplace-Operator on Riemannian Manifolds, *Canad. J. Math.* **1** (1949) pp.242–256.

[N] Y. Nambu: Quark Model and the Factorization of the Veneziano Amplitude, in "Symmetries and Quark Models", Ed. by R. Chand, Gordon and Breach (1970) pp.269–278.

[Polc] J. Polchinski: Evaluation of the One-Loop String Path Integral, *Comm. Math. Phys.* **104** (1986) pp.37–47.

[Poly] A. M. Polyakov: Quantum Geometry of Bosonic Strings, *Phys. Lett.* **103B** (1981) pp.207–210.

[Ra] P. Ramond: *Field Theory, A modern Primer*, Benjamin-Cummings (1981).

[RS] D. B. Ray and I. M. Singer: Analytic Torsion for Complex Manifolds, *Ann. Math.* **98** (1973) pp.154–177.

[Ry] L. H. Ryder: *Quantum Field Theory*, Cambridge University Press (1985).

[Sche] J. Scherk: An Introduction to the Theory of Dual Models and Strings, *Rev. Mod. Phys.* **47** (1975) pp.123–164.

[SchwarzA] A. S. Schwarz: Instantons and Fermions in the Field of Instanton, *Comm. Math. Phys.* **64** (1979) pp.233–268.

[SchwarzJ] J. H. Schwarz: Superstring Theory, *Phys. Reports* **89** (1982) pp.223–322.

[See1] R. Seeley: Complex Powers of an Elliptic Operator, in A.M.S. Proc. Symp. Pure Math. no.**10** (1967) pp.288–307.

[See2] R. Seeley: The Resolvent of an Elliptic Boundary Problem, *Am. J. Math.* **91** (1969) pp.889–921.

[See3] R. Seeley: Analytic Extension of the Trace Associated with Elliptic Boundary Problems, *Am. J. Math.* **91** (1969) pp.963–983.

[Seg1] G. Segal: Unitary Representations of some Infinite Dimensional Groups, *Comm. Math. Phys.* **80** (1981) pp.301–342.

[Seg2] G. Segal: *The Definition of Conformal Field Theory*, preprint (1987).

[Ser] J-P. Serre: *A Course in Arithmetic*, Springer (1973).

[Sie] C. L. Siegel: *Lectures on Advanced Analytic Number Theory*, Tata Institute (1961).

[Sim] B. Simon: *Functional Integration and Quantum Physics*, Academic Press (1979).

[Su] L. Susskind: Structures of Hadrons Implied by Duality, *Phys. Rev.* **D1** (1970) pp.1182–1186.

[TK] A. Tsuchiya and Y. Kanie: Vertex Operators in Conformal Field Theory on $\mathbb{P}^1$ and Monodromy Representations of Braid Group, in "Conformal Field Theory and Solvable Lattice Models", Advanced Studies in Pure Mathematics **16** (1988) pp.297–372.

[V] M. A. Virasoro: Subsidiary Conditions and Ghosts in Dual-Resonance Models, *Phys. Rev.* **D1** (1970) pp.2933–2936.

[W1] E. Witten: Physics and Geometry, in Proceedings of the International Congress of Mathematicians, Berkeley, California, USA, 1986 (1987) pp.267–303.

[W2] E. Witten: Quantum Field Theory, Grassmannians, and Algebraic Curves, *Comm. Math. Phys.* **113** (1988) pp.529–600.

Haruo Tsukada

Department of Mathematics

University of California, San Diego

La Jolla, California 92093

MEMOIRS of the American Mathematical Society

SUBMISSION. This journal is designed particularly for long research papers (and groups of cognate papers) in pure and applied mathematics. The papers, in general, are longer than those in the TRANSACTIONS of the American Mathematical Society, with which it shares an editorial committee. Mathematical papers intended for publication in the Memoirs should be addressed to one of the editors:

Ordinary differential equations, partial differential equations and applied mathematics to ROGER D. NUSSBAUM, Department of Mathematics, Rutgers University, New Brunswick, NJ 08903

Harmonic analysis, representation theory and Lie theory to AVNER D. ASH, Department of Mathematics, The Ohio State University, 231 West 18th Avenue, Columbus, OH 43210

Abstract analysis to MASAMICHI TAKESAKI, Department of Mathematics, University of California, Los Angeles, CA 90024

Real and harmonic analysis to DAVID JERISON, Department of Mathematics, M.I.T., Rm 2–180, Cambridge, MA 02139

Algebra and algebraic geometry to JUDITH D. SALLY, Department of Mathematics, Northwestern University, Evanston, IL 60208

Geometric topology and general topology to JAMES W. CANNON, Department of Mathematics, Brigham Young University, Provo, UT 84602

Algebraic topology and differential topology to RALPH COHEN, Department of Mathematics, Stanford University, Stanford, CA 94305

Global analysis and differential geometry to JERRY L. KAZDAN, Department of Mathematics, University of Pennsylvania, E1, Philadelphia, PA 19104-6395

Probability and statistics to RICHARD DURRETT, Department of Mathematics, Cornell University, Ithaca, NY 14853-7901

Combinatorics and number theory to CARL POMERANCE, Department of Mathematics, University of Georgia, Athens, GA 30602

Logic, set theory, general topology and universal algebra to JAMES E. BAUMGARTNER, Department of Mathematics, Dartmouth College, Hanover, NH 03755

Algebraic number theory, analytic number theory and modular forms to AUDREY TERRAS, Department of Mathematics, University of California at San Diego, La Jolla, CA 92093

Complex analysis and nonlinear partial differential equations to SUN-YUNG A. CHANG, Department of Mathematics, University of California at Los Angeles, Los Angeles, CA 90024

All other communications to the editors should be addressed to the Managing Editor, DAVID J. SALTMAN, Department of Mathematics, University of Texas at Austin, Austin, TX 78713.